Memoirs of the American Mathematical Society
Number 210

Peter H. Krauss and David M. Clark

Global subdirect products

Published by the
AMERICAN MATHEMATICAL SOCIETY
Providence, Rhode Island, USA

January 1979 · Volume 17 · Number 210 (first of 3 numbers)

TABLE OF CONTENTS

ABSTRACT

An internal characterization is given of those subdirect products which are structures of global sections of discrete sheaves. Such subdirect products are called global. Patching of subdirect products over a dual ring of subsets of the index set is defined, and a uniform method of constructing global subdirect products from the patching property is developed. The role of the hull-kernel topology in sheaf constructions is analysed in the setting of universal algebra. Global subdirect products which come from Hausdorff sheaves over Boolean spaces (Boolean subdirect products) are treated in terms of the normal transform. Global representation of varieties is defined and investigated. Finally, applications to the sheaf representation of rings and lattice ordered rings are given.

AMS (MOS) subject classifications(1970) : Primary 08A05, 08A25. Secondary 08A15,16A48,06A70.

Key words and phrases : Sheaf representation, patchwork property, equalizer topology, hull-kernel topology, structure of global sections, Boolean subdirect product, normal transform, representation of varieties, semi-prime rings, biregular rings, Baer rings, lattice-ordered rings, function rings.

Library of Congress Cataloging in Publication Data CIP

Krauss, Peter H., 1933-
Global subdirect products.

(Memoirs of the American Mathematical Society ; no. 210)
Bibliography: p.
1. Algebra, Universal. 2. Associative rings. 3. Sheaves, Theory of. I. Clark, David M., 1947- joint author. II. Title. III. Series: American Mathematical Society. Memoirs ; no. 210.
QA3.A57 no. 210 [QA251] 510'.8s [512] 78-23373
ISBN 0-8218-2210-1

GLOBAL SUBDIRECT PRODUCTS

A celebrated theorem of Birkhoff asserts that every (non-trivial) algebra is isomorphic to a subdirect product of subdirectly irreducible algebras. Why is this not the ultimate representation theorem of (universal) algebra? There are two reasons for this. The basic idea behind representation theorems is to "decompose" a given structure into "simpler" structures in such a way that the properties of the given structure can be "reduced" to the properties of the simpler structures. In the case of Birkhoff's Theorem the "simpler" structures are subdirectly irreducible and the "decomposition" is subdirect. Now, first of all it is usually very difficult to determine the subdirectly irreducible factors (that is the subdirectly irreducible homomorphic images) of a given algebra. In fact, frequently these factors are just not known at all (as is the case,for example, for arbitrary groups or rings). However, there are important cases where the subdirect factors are known. For example, an abelian group is subdirectly irreducible if and only if it is cocyclic (that is quasi cyclic or cyclic of prime power order). Thus we obtain: Every (non-trivial) abelian group is isomorphic to a subdirect product of cocyclic groups. Although the properties of cocyclic groups are very well known, this theorem plays practically no role in abelian group theory. This is due to the second weakness of Birkhoff's Theorem: Subdirect products in general are so "loose" that very little can be inferred from the properties of the factors. In the structure theory of abelian groups the central role is played by <u>direct sum representations</u>. Direct sums are special subdirect products which are "tight" enough so that significant information can be obtained from the properties of the factors.

To further illustrate this weakness of Birkhoff's Theorem, let us consider another subdirect representation by subdirectly irreducibles: A (non-trivial) commutative ring is semi-simple (that is has trivial Jacobson radical) if and only if it is a subdirect product of fields. Although this is an important result of ring theory (in fact it motivated the definition of

---

Received by the Editors July 27,1977.

The second author was partially supported for this work at the Gesamthochschule Kassel by a grant from the Alexander von Humboldt Foundation of West Germany.

the Jacobson radical) it again does not tell us that much about semi-simple commutative rings, and ring theorists are known to complain about this shortcoming. Although direct sums of rings exist, interesting direct sum representations for important classes of rings apparently don't. This dilemma appears to have provided a major motivation for the development of <u>sheaf representations</u>. Structures of global sections of sheaves are again special subdirect products which are "tight" enough to allow significant conclusions to be drawn from the properties of the factors (or stalks).

Both in the case of direct sum representations as well as in the case of sheaf representations the crucial issue of course is to explicate the underlying notion of "tightness" which renders these constructions interesting. Now we shall see that direct sum representations are sheaf representations in a perfectly natural and direct sense, so that from this point of view the sheaf construction is a natural generalization of the direct sum construction. However, what exactly is it that makes these subdirect representations "tight"? It appears that this question has never been explicitly pursued in the literature. Sheaf representations apparently "just have nice properties". Implicitly the so-called "patchwork property" plays a central role in sheaf representation. Structures of global sections patch over the equalizer topology, and it is plausible that this accounts for the "tightness" of the subdirect representation.

Unrestricted patching over the equalizer topology actually characterizes structures of global sections, as was noticed independently by Weispfenning [38]. However, unrestricted patching is a condition which is difficult to verify in applications. In fact, in applications unrestricted patching is usually reduced to finite patching via some kind of compactness argument. Therefore we first introduce the more general notion of (finite) <u>patching over a dual ring of subsets</u> of the index set of a subdirect product, and then we give a uniform procedure for "tightening" a subdirect product which patches finitely to a sheaf representation which patches unrestrictedly. This procedure is called <u>globalization</u>. It is carried out in the setting of universal algebra and applies to all sheaf constructions in algebra which start from finite patching. Many discrete sheaf constructions in algebra <u>directly</u> start from finite patching and it appears to us that <u>all</u> discrete sheaf constructions in algebra start from finite patching in some way or other. We shall amply support this view with examples. As a practical consequence of our analysis many major papers on the sheaf construction can be substantially shortened and simplified with our methods, in fact some of them collapse to triviality. Moreover, in many instances of sheaf constructions known from the literature we shall show that <u>universal</u> algebra yields all the information the authors obtain in the special setting of some particular algebra, and in many other instances we shall show that this information consists of universal algebra in conjunction with a minute amount of special algebra.

Now in this process of globalization frequently something rather peculiar is happening. As mentioned above, the main shortcomings of Birkhoff's Theorem are twofold: First, too little is known about the factors of the subdirect representation and secondly, the subdirect product is not "tight" enough to establish a useful connection to the factors. In our view one deficiency is just as serious as the other. From this point of view it is interesting to notice that many sheaf constructions actually start from a subdirect representation where a good deal is known about the factors and where the factors are indeed "simpler" than the subdirect product in some tangible and algebraically meaningful sense (e.g. a subdirect product of prime rings). However, the subdirect representation is not deemed "tight" enough. To obtain a sheaf representation the crucial step is to establish that the given subdirect representation patches over a suitable dual ring of subsets of the index set (e.g. the dual ring of hull-kernel open sets). If equalizers are open and the index space is compact (with respect to the topology whose basis is the dual ring) then such a subdirect product is essentially already a structure of global sections of a sheaf. This is well-known, however there is a fly in the oatmeal: Frequently equalizers are not open (e.g. in the hull-kernel topology) or the index space is not compact (e.g. with the equalizer topology) or both. In these cases the typical sheaf construction amounts to a simple trade-off: The subdirect representation is "tightened" at the expense of almost completely losing track of the factors. It turns out that this process of "tightening" the subdirect representation can be uniformly described in the setting of universal algebra as globalization. Moreover, in this setting it will become clear how in this process the factors usually are blown up and proliferated into obscurity. This accounts for the striking fact (easily accessible in the literature) that many architects of sheaf constructions know practically nothing about the factors (or stalks) of the subdirect representation they erect. We rather doubt the merits of this procedure because it appears to us that the algebraically interesting information comes directly from the original subdirect representation (which is "tight" in a somewhat less rigid but still rather significant fashion and where something substantial is known about the factors) and not from the final sheaf representation. We shall more explicitly discuss this point as we develop our exposition in detail.

Sheaves of structures are rather complicated mathematical constructions which involve two topological spaces and a local homeomorphism. However, in applications to algebra sheaves are only used as tools to construct a subdirect product, the structure of global sections of the sheaf. We shall define a very simple closure condition for subdirect products and call a subdirect product global if it is closed in this respect. Subsequently we prove the following:

Theorem The structures of global sections of sheaves are exactly the global subdirect products of disjoint structures.

This characterization should completely dispense with those sheaf constructions in algebra whose only purpose is a global subdirect representation, because it turns out that in these cases one always can verify directly that some natural subdirect representation is global without rigging up the cumbersome apparatus of a sheaf. This method by itself tends to simplify matters considerably. Moreover it facilitates an analysis of the peculiar "tightness" of global subdirect products which leads to the notion of patching over a dual ring of subsets of the index set. This notion is at the core of all sheaf representation. The rest is universal algebra.

Of course there are several well-known attempts in the literature to analyse the sheaf construction in the setting of universal algebra (see, e.g., Davey [13] , Comer [8], Wolf [41] and Burris and Werner [5]). However, each of these explications has only limited applications to (admittedly large classes of) special sheaf constructions known from the literature. There are other authors who substantially contribute to the universal algebra of the sheaf construction (see,e.g. Kennison [26], Ledbetter [29] and Werner [40]). Weispfenning [38] analyzes the sheaf construction in a model theoretic setting and obtains results which are apparently closely related to ours. However, Weispfenning's work has been carried out simultaneously with ours and has only come to our attention after the completion of this manuscript. In general we make no attempt to give an historically complete account of the sheaf construction in algebra, or even to take into consideration every contribution in this area. Our main goal is to present a unifying approach to the sheaf construction in the setting of universal algebra. This approach is based on a few simple ideas which we have extracted from the literature. The main purpose of this paper is to demonstrate the universality of these ideas. This also has guided our selection of applications. We have inspected far more examples in the literature than we explicitly discuss in this study, or even than we mention in the references. We have found none to which our methods do not apply. Out of these examples we have chosen a few for actual presentation which are generally considered significant results in sheaf representation and which are characteristic of the ideas and methods involved in the sheaf construction. Finally, we have totally ignored some alternate approaches to the sheaf construction, notably the category and topos oriented approaches. This sufficiently illustrates the limitations of this study.

In Section 1 we fix notation and terminology and review a few well-known facts which will be needed later. In Section 2 we characterize subdirect products which arise from sheaf constructions and treat some well-known algebraic constructions in this framework, such as direct sums and Boolean powers. Subsequently we convert some basic notions connected with the sheaf construction into a form which is more suitable to our setting. In Section 3 we investigate the hull-kernel topology in the setting of universal algebra. It turns out that many sheaf constructions involve in some form or other the hull-kernel topology induced by a subdirect representation. Section 4 contains the main results of this study. We give a uniform method of constructing global subdirect products from the patching property. In Section 5 we consider subdirect products which come from Hausdorff sheaves over Boolean spaces. Such subdirect products are called *Boolean*, and we describe the special role played by the normal transform in this setting. In Section 6 we pose the problem of global subdirect representation of varieties and discuss some partial solutions to the problem. In particular, we give Kennison's representation of *function rings* [26], Ledbetter's representation of *vector groups* and *relative Stone algebras* (written communication) and Bulman-Fleming and Werner's representation of *discriminator varieties* [4] . In Sections 7 and 8 we give examples from ring theory where sheaf representation has been most successful. We have selected a representative sample from the vast literature in this area, including Hofmann's results on *semi-prime rings with identity* [20], Dauns and Hofmann's results on *biregular rings* [11] and *weakly biregular rings* [12], Hofmann's results on *Baer rings* [20], Kennison's results on global subdirect representation by *integral domains* [26] and Keimel's results on *lattice-ordered rings* [24].

1. PRELIMINARIES

In this section we shall first make a few remarks on notation and terminology, and then we shall review some well-known facts for the purpose of reference. Given is a similarity type determined by a set Rl of relation symbols and a set Op of operation symbols. A structure $\mathfrak{A}$ has universe $A = |\mathfrak{A}|$, and for each n-ary relation symbol $R \in Rl$ has an n-ary relation $R^{\mathfrak{A}}$, and for each n-ary operation symbol $f \in Op$ has an n-ary operation $f^{\mathfrak{A}}$. $\mathfrak{A}$ is called an algebra if $Rl = \emptyset$ and $\mathfrak{A}$ is called a relational structure if $Op = \emptyset$.

If $\mathfrak{M}$ is a class of structures then $I\mathfrak{M}$, $S\mathfrak{M}$ and $H\mathfrak{M}$ denote the classes of isomorphic images, substructures and homomorphic images of members of $\mathfrak{M}$ respectively, and $P\mathfrak{M}$, $P_s\mathfrak{M}$, $P_r\mathfrak{M}$ and $P_u\mathfrak{M}$ denote the classes of direct products, subdirect products, reduced direct products, and ultraproducts of subsets of $\mathfrak{M}$ respectively. We allow the direct product of the empty set of structures.

Next we introduce the (finitary) first-order language of the given similarity type. Then $A\mathfrak{M}$, $O\mathfrak{M}$ and $E\mathfrak{M}$ denote the atomical, universal and elementary classes generated by $\mathfrak{M}$ respectively. For a class $\mathfrak{M}$ of algebras, $A\mathfrak{M}$ is called the variety generated by $\mathfrak{M}$ and is denoted by $V\mathfrak{M}$.

For any structure $\mathfrak{A}$, $0^{\mathfrak{A}}$ is the trivial congruence on $\mathfrak{A}$ and $1^{\mathfrak{A}}$ is the universal congruence on $\mathfrak{A}$.

Let $\mathfrak{B} \subseteq \Pi\langle \mathfrak{A}_i \mid i \in I\rangle$. For each $F \subseteq I$, π_F is the projection of $\Pi\langle \mathfrak{A}_i \mid i \in I\rangle$ onto $\Pi\langle \mathfrak{A}_i \mid i \in F\rangle$, and θ_F and $\theta_F^{\mathfrak{B}}$ are the projection congruences on $\Pi\langle \mathfrak{A}_i \mid i \in I\rangle$ and $\mathfrak{B}$ induced by π_F respectively. $\mathfrak{B}$ is called non-trivially subdirect if $\mathfrak{B}$ is subdirect and for no $i \in I$ is $\pi_i : \mathfrak{B} \to \mathfrak{A}_i$ an isomorphism. If $x, y \in B$ then $\theta^{\mathfrak{B}}(x, y)$ denotes the smallest congruence χ on $\mathfrak{B}$ such that $x\chi y$. A congruence χ on $\mathfrak{B}$ is called principal if there exist $x, y \in B$ such that $\chi = \theta^{\mathfrak{B}}(x, y)$. If α is an ordinal and $x \in (\Pi\langle A_i \mid i \in I\rangle)^\alpha$ then $\pi_F(x) \in (\Pi\langle A_i \mid i \in F\rangle)^\alpha$ is defined by

$$\pi_F(x)(\xi) = \pi_F(x(\xi)).$$

If φ is a formula and $x \in A^\omega$ then $\mathfrak{A} \models \varphi[x]$ means that x satisfies φ in $\mathfrak{A}$, and we adopt the usual conventions to make satisfaction of formulas by finite sequences well-behaved. If $x \in (\Pi A_i)^n$ then we define

$$[\![\varphi(x)]\!] = \{i \in I \mid \mathfrak{A}_i \models \varphi[\pi_i(x)]\}$$

and for $x, y \in \Pi A_i$ we write

$$E(x, y) = [\![x = y]\!], \quad D(x, y) = [\![x \neq y]\!].$$

The <u>ternary discriminator operation</u> and the <u>normal transform</u> are defined by

$$t(x, y, z) = \begin{cases} z & \text{if } x = y \\ x & \text{if } x \neq y \end{cases}$$

$$n(x, y, z, w) = \begin{cases} z & \text{if } x = y \\ w & \text{if } x \neq y \end{cases}$$

respectively. t and n canonically induce operations $\bar{t}$ and $\bar{n}$ on $\Pi\langle A_i \mid i \in I\rangle$:

$$\bar{t}(x, y, z)(i) = t(x(i), y(i), z(i))$$

$$\bar{n}(x, y, z, w)(i) = n(x(i), y(i), z(i), w(i))$$

Our first topic for review is the <u>Wallman-compactification</u> of a topological space. This compactification is constructed from the ring of closed sets. The details are well-known and the pertinent facts are easy to prove (Wallman [36]). Let I be a non-empty set and let $\mathcal{R}$ be a family of subsets of I. $\mathcal{R}$ is called a <u>ring of subsets</u> of I if

(i) $\emptyset \in \mathcal{R}$;

(ii) if $F, G \in \mathcal{R}$ then $F \cup G, F \cap G \in \mathcal{R}$.

A ring of subsets of I is called <u>Boolean</u> if

(iii) if $F, G \in \mathcal{R}$ then $F - G \in \mathcal{R}$,

and a Boolean ring of subsets of I is called a <u>field</u> if

(iv) $I \in \mathcal{R}$.

We say that $\mathcal{R}$ <u>covers</u> I if $\bigcup \mathcal{R} = I$ and we define

$$\breve{\mathcal{R}} = \{I - F \mid F \in \mathcal{R}\}.$$

Let $\mathcal{R}$ be a ring of subsets of I and let J and M be the sets of prime filters and maximal filters on $\mathcal{R}$ respectively. Since $\mathcal{R}$ is a distributive lattice, $M \subseteq J$. For each $i \in I$ define

$$h(i) = \{F \in \mathcal{R} \mid i \in F\}.$$

Since $\emptyset \in \mathcal{R}$, $h(i) \neq \mathcal{R}$. Thus for every $i \in I$, $h(i) = \emptyset$ or $h(i) \in J$. Moreover, $\mathcal{R}$ covers I if and only if for every $i \in I$, $h(i) \in J$. For each

$F \in \mathcal{R}$ define

$$h^*(F) = \{\mathcal{U} \in J \mid F \in \mathcal{U}\}$$

and let

$$\mathcal{R}^* = \{h^*(F) \mid F \in \mathcal{R}\} .$$

Then $\mathcal{R}^*$ is a ring of subsets of J and $h^*: \mathcal{R} \to \mathcal{R}^*$ is an isomorphism. Since $\emptyset \in \mathcal{R}$, $\mathcal{R}$ is a basis of closed sets for a topology $\mathcal{T}(\mathcal{R})$ on I and $\mathcal{R}^*$ is a basis of closed sets for a topology $\mathcal{T}(\mathcal{R}^*)$ on J.

Lemma 1.1 J is a compact T_0-space. ∎

Lemma 1.2 For every $F \in \mathcal{R}$,

(i) $h(F) = h(I) \cap h^*(F)$

(ii) $h(I - F) = h(I) - h^*(F)$. ∎

Corollary 1.3 $h(I) \cap J$ is a dense subset of J. ∎

Lemma 1.4 If $\mathcal{R}$ covers I then the following are equivalent:

(i) I is a T_0-space.

(ii) h is one-to-one.

(iii) h is a homeomorphism of I onto $h(I)$ (with the relative topology). ∎

If $\mathcal{R}$ covers I and I is a T_0-space then J is called the Wallman T_0-compactification of I.

Lemma 1.5 I is compact if and only if $M \subseteq h(I)$. ∎

Corollary 1.6 If $h(I) = J$ then I is compact. ∎

Lemma 1.7 Every open subset of I is the union of closed sets if and only if $h(I) \subseteq M$. ∎

Notice, if I is a T_1-space then every (open) subset of I is the union of closed sets, and if every open subset of I is the union of closed sets then $\mathcal{R}$ covers I.

<u>Corollary</u> 1.8 Suppose every open subject of I is the union of closed sets. Then

(i) $h(I)$ is a dense subset of M;

(ii) I is compact if and only if $h(I) = M$. ∎

<u>Lemma</u> 1.9 M is a compact T_1-space (with the relative topology). ∎

If I is a T_1-space then M is called the <u>Wallman</u> <u>T_1-compactification</u> of I.

Next we shall quickly compile some information for the case where $\mathcal{R}$ is a <u>Boolean</u> ring. In this case I is a T_0-space if and only if I is a T_1-space. Moreover, $M = J$. If

$$\mathcal{F} = \mathcal{R} \cup \breve{\mathcal{R}}$$

then $\mathcal{F}$ is a field of subsets of I and $\mathcal{R} = \mathcal{F}$ if and only if $I \in \mathcal{R}$.

<u>Lemma</u> 1.10 If $I \notin \mathcal{R}$ then

(i) $\breve{\mathcal{R}}$ is a prime filter on $\mathcal{F}$;

(ii) for every prime filter $\mathcal{U}$ on $\mathcal{R}$ there exists a unique prime filter $\mathcal{V}$ on $\mathcal{F}$ such that $\mathcal{U} = \mathcal{V} \cap \mathcal{F}$; moreover, $\mathcal{V} \neq \breve{\mathcal{R}}$;

(iii) for every prime filter $\mathcal{V}$ on $\mathcal{F}$, if $\mathcal{V} \neq \breve{\mathcal{R}}$ then $\mathcal{V} \cap \mathcal{R}$ is a prime filter on $\mathcal{R}$. ∎

Now there are two topologies on I, $\mathfrak{T}(\mathcal{R})$ and $\mathfrak{T}(\mathcal{F})$. We shall discuss the space $(I, \mathfrak{T}(\mathcal{F}))$ in case the space $(I, \mathfrak{T}(\mathcal{R}))$ is compact.

<u>Lemma</u> 1.11 Suppose $(I, \mathfrak{T}(\mathcal{R}))$ is compact. Then $(I, \mathfrak{T}(\mathcal{F}))$ is compact if and only if $I \in \mathcal{R}$ or $\mathcal{R}$ does not cover I.

<u>Proof</u>: If $(I, \mathfrak{T}(\mathcal{F}))$ is compact and $\mathcal{R}$ covers I then clearly $I \in \mathcal{R}$. Conversely, suppose $\mathcal{R}$ does not cover I and $\mathcal{C} \subseteq \mathcal{F}$ has the finite intersection property. We distinguish two cases. <u>Case</u> 1 $\mathcal{C} \cap \mathcal{R} \neq \emptyset$. Pick $F \in \mathcal{C} \cap \mathcal{R}$. Then

$$\bigcap \mathcal{C} = \bigcap \{F \cap G \mid G \in \mathcal{C}\}$$

Since $\mathcal{R}$ is Boolean, for every $G \in \mathcal{F}$, $F \cap G \in \mathcal{R}$. Since $(I, \mathfrak{T}(\mathcal{R}))$ is compact, $\bigcap \mathcal{C} \neq \emptyset$. <u>Case</u> 2 $\mathcal{C} \cap \mathcal{R} = \emptyset$. Then $\mathcal{C} \subseteq \breve{\mathcal{R}}$ and $\bigcap \mathcal{C} \neq \emptyset$ because $\bigcap \breve{\mathcal{R}} \neq \emptyset$. This establishes that $(I, \mathfrak{T}(\mathcal{F}))$ is compact. ∎

Lemma 1.12 Suppose $(I, \mathfrak{T}(\mathcal{R}))$ is compact whereas $(I, \mathfrak{T}(\mathfrak{F}))$ is not. If J is the set of prime filters on $\mathfrak{F}$ then

$$J = h(I) \cup \{\breve{\mathcal{R}}\}.$$

Proof: Since $I \notin \mathcal{R}$, by Lemma 1.10, $\breve{\mathcal{R}} \in J$. Suppose $\mathcal{V} \in J$, where $\mathcal{V} \neq \breve{\mathcal{R}}$ and let $\mathcal{U} = \mathcal{V} \cap \mathcal{R}$. By Lemma 1.10, $\mathcal{U}$ is a prime filter on $\mathcal{R}$. Since $(I, \mathfrak{T}(\mathcal{R}))$ is compact, $\bigcap \mathcal{U} \neq \emptyset$. Let $i \in \bigcap \mathcal{U}$. By Lemma 1.10, $\mathcal{V} = h(i)$. ∎

Our second topic for review are redundancies in subdirect products. We shall be concerned with redundancies which play a central role in the investigation of quasi primal algebras (See Krauss [27]). Again let $\mathcal{B} \subseteq \Pi\langle \mathfrak{A}_i \mid i \in I\rangle$ be subdirect. For $i, j \in I$ we define $i \sim j(\mathcal{B})$ if there exists an isomorphism f from $\mathfrak{A}_i$ onto $\mathfrak{A}_j$ such that for all $x \in B$, $f(x(i)) = x(j)$. $J \subseteq I$ is called a $\mathcal{B}$-transversal if for every $i, j \in J$, $i \sim j(\mathcal{B})$ if and only if $i = j$. J is called $\mathcal{B}$-complete if for every $i \in I$ there exists $j \in J$ such that $i \sim j(\mathcal{B})$. $\mathcal{B}$ is called irredundant if I is a $\mathcal{B}$-transversal.

Lemma 1.13 If $\mathcal{B}$ is an algebra then for every $i, j \in I$, $i \sim j(\mathcal{B})$ if and only if $\theta_i^{\mathcal{B}} = \theta_j^{\mathcal{B}}$. ∎

Lemma 1.14 If J is a complete $\mathcal{B}$-transversal then

$$\pi_J : \mathcal{B} \to \Pi\langle \mathfrak{A}_j \mid j \in J\rangle$$

is a subdirect embedding and $\pi_J(\mathcal{B})$ is irredundant. ∎

Lemma 1.14 tells us that we can always "make a subdirect product irredundant" by "removing redundancies". We shall see that many special properties of subdirect representations we are interested in are preserved by this process.

2. GLOBAL SUBDIRECT PRODUCTS

As we pointed out in the introduction, the main purpose of sheaf constructions in algebra is the construction of special subdirect products, namely the structures of global sections of sheaves. The special algebraic properties of such subdirect products are then deduced from the special set theoretic features of this construction. Although familiar to the specialist, the sheaf construction involves considerable notational and conceptual machinery which tends to render access to the relevant algebraic facts cumbersome and involved. If only to "fix notation and terminology", this machinery is developed in the literature over and over again. In contrast, we shall give a simple internal characterization (Theorem 2.28) of subdirect products which are structures of global sections of discrete sheaves. In applications this condition can be directly verified without erecting the elaborate edifice of a sheaf. Moreover, the relevant algebraic features of structures of global sections can all be expressed in terms of this internal characterization. In our view the advantages of this procedure appear to be considerable: The application of this method alone would reduce the voluminous literature on sheaf representation in algebra by an appreciable amount. Of course, there are important applications of the sheaf construction which go far beyond the "subdirect representation aspect". We only consider the sheaf construction as a tool to construct special subdirect representations of algebras. We claim that this tool can be replaced by a much simpler and more directly applicable device. In this section we shall establish this claim.

Traditionally algebraists consider structures with operations only, that is algebras. The consideration of structures with relations has entered the subject through the "model theoretic" approach. In fact, from the model theoretic point of view, which is characterized by the consideration of "formulas", sometimes structures with relations only, that is relational structures, are regarded more natural. Although by now lively communication between the two camps has developed, the two approaches have never really fused. One of the main reasons for this situation appears to be the fact that relations behave peculiarly with respect to homomorphisms, and what would an algebraist do without the familiar properties of his favorite morphisms?

Accordingly, the bulk of the literature on subdirect representation of structures by global sections concerns algebras. In fact, model theorists introduced structures with relations rather belatedly into the development of sheaf theory, and then their treatment sometimes appears to be somewhat ad hoc. Nevertheless, model theorists have made important contributions to sheaf representation theory (see, e.g., Comer [10], Macintyre [30], Weispfenning [37], Burris and Werner [5]), so that it appears necessary at least initially to treat representation by structures of global sections of sheaves in a framework which takes relations into account. We shall do this. Unfortunately this may irritate some algebraists who don't like this "odd formula business". To accomodate these people we should like to point out that algebraists are actually quite familiar with formulas, namely equations and inequations (that is negations of equations). However, if $\mathcal{B} \subseteq \Pi\langle \mathfrak{A}_i \mid i \in I\rangle$ and $\sigma = \tau$ is an equation then for any $x \in B^n$ there exist $y, z \in B$ such that

$$[\![(\sigma = \tau)(x)]\!] = E(y, z)$$

$$[\![(\sigma \neq \tau)(x)]\!] = D(y, z).$$

Thus if the language has no relation symbols then

$$\{[\![\varphi(x)]\!] \mid \varphi \text{ atomic}, x \in B^n, n<\omega\} = \{E(y, z) \mid y, z \in B\}$$

$$\{[\![\neg\varphi(x)]\!] \mid \varphi \text{ atomic}, x \in B^n, n<\omega\} = \{D(y, z) \mid y, z \in B\}$$

so that one can read $[\![\varphi(x)]\!]$ and $[\![\neg\varphi(x)]\!]$, where φ is an atomic formula, as $E(y, z)$ and $D(y, z)$ respectively, and one never has to mention formulas at all. In the more special sections of this study (Sections 3,5,6,7 and 8) we shall retreat to this position anyway. However, in this section and in Section 4 we shall include the more general situation where some atomic formulas come from relation symbols so that the reduction to equalizers is not possible any more.

Let $\mathcal{B} \subseteq \Pi\langle \mathfrak{A}_i \mid i \in I\rangle$. The <u>equalizer topology on</u> I <u>induced by</u> $\mathcal{B}$ is the topology on I with subbasis consisting of all sets of the form $[\![\varphi(x)]\!]$, where φ is an atomic formula and $x \in B^n$.

We should like to point out that usually these sets do not form a basis for a topology. Moreover, in most applications it is quite difficult to get a hold on the equalizer topology unless it is explicitly related to some algebraic properties of $\mathcal{B}$. Nevertheless, this topology plays the central role in sheaf constructions.

Let $\mathfrak{T}$ be a topology on I containing the equalizer topology induced by $\mathcal{B}$. For $x \in \Pi A_i$ we define

$$x \in \Gamma(B, \mathfrak{T}) \text{ iff } x \in \Pi\pi_i(B) \text{ and for every } y \in B,\ E(x, y) \in \mathfrak{T}.$$

<u>Lemma</u> 2.1 For each $f \in Op$, $x \in \Gamma(B, \mathfrak{T})^n$ and $y \in B$,

$$E(f(x), y) = \bigcup_{u \in B^n} [\bigcap_{k<n} E(x_k, u_k) \cap E(f(u), y)]$$

<u>Proof</u>: Since $x \in (\Pi \pi_i(B))^n$,

$$f(x_o(i), \ldots, x_{n-1}(i)) = y(i)$$

if and only if for some $u \in B^n$, $\bigwedge_{k<n} x_k(i) = u_k(i)$ and

$$f(u_o(i), \ldots, u_{n-1}(i)) = y(i). \blacksquare$$

<u>Corollary</u> 2.2 $B \subseteq \Gamma(B, \mathfrak{T}) \subseteq \Pi\langle\pi_i(B) \mid i \in I\rangle$ and $\Gamma(B, \mathfrak{T})$ is closed under the operations of $\Pi \mathfrak{A}_i$. $\blacksquare$

Let $\Gamma(\mathfrak{B}, \mathfrak{T})$ be the substructure of $\Pi \mathfrak{A}_i$ determined by $\Gamma(B, \mathfrak{T})$. $\mathfrak{B}$ is called a $\mathfrak{T}$-<u>global substructure</u> of $\Pi\langle \mathfrak{A}_i \mid i \in I\rangle$ if $\mathfrak{B} \subseteq \Pi\langle \mathfrak{A}_i \mid i \in I\rangle$, $\mathfrak{T}$ is a topology on I containing the equalizer topology induced by $\mathfrak{B}$ and $\mathfrak{B} = \Gamma(\mathfrak{B}, \mathfrak{T})$.

<u>Corollary</u> 2.3 Suppose $\mathfrak{B} \subseteq \Pi\langle \mathfrak{A}_i \mid i \in I\rangle$ and $\mathfrak{T}$ is a topology on I. Then $\mathfrak{B}$ is $\mathfrak{T}$-global if and only if

(i) for every atomic formula φ and every $x \in B^n$, $[\![\varphi(x)]\!] \in \mathfrak{T}$;

(ii) for every $x \in \Pi\pi_i(B)$, if $E(x, y) \in B$ for all $y \in B$, then $x \in B$. $\blacksquare$

<u>Lemma</u> 2.4 Suppose $\mathfrak{B} \subseteq \Pi\langle \mathfrak{A}_i \mid i \in I\rangle$. Then for every atomic formula φ and every $x \in (\Pi\pi_i(B))^n$,

$$[\![\varphi(x)]\!] = \bigcup_{x \in B^n} [\bigcap_{k<n} E(x_k, y_k) \cap [\![\varphi(y)]\!]]. \blacksquare$$

<u>Corollary</u> 2.5 Suppose $\mathfrak{B} \subseteq \Pi\langle \mathfrak{A}_i \mid i \in I\rangle$ and $\mathfrak{T}$ is a topology on I. If $\mathfrak{T}$ contains the equalizer topology induced by $\mathfrak{B}$, then $\mathfrak{T}$ contains the equalizer topology induced by $\Gamma(\mathfrak{B}, \mathfrak{T})$ and

$$\Gamma(\Gamma(\mathfrak{B}, \mathfrak{T}), \mathfrak{T}) = \Gamma(\mathfrak{B}, \mathfrak{T}),$$

i.e. $\Gamma(\mathfrak{B}, \mathfrak{T})$ is a $\mathfrak{T}$-global substructure of $\Pi\langle \mathfrak{A}_i \mid i \in I\rangle$.

<u>Proof</u>: The first part of the assertion follows immediately from Lemma 2.4. In particular, the second part of the assertion makes sense. To prove it, suppose $x \in \Gamma(\Gamma(B, \mathfrak{T}), \mathfrak{T})$. Then for every $y \in \Gamma(B, \mathfrak{T})$, $E(x, y) \in \mathfrak{T}$. Since $B \subseteq \Gamma(B, \mathfrak{T})$, $x \in \Gamma(B, \mathfrak{T})$. $\blacksquare$

If $\mathcal{B} \subseteq \Pi\langle \mathfrak{A}_i \mid i \in I\rangle$ then by Corollary 2.2, $\Gamma(\mathcal{B}, \mathfrak{T})$ "fills in" $\Pi\langle \pi_i(\mathcal{B}) \mid i \in I\rangle$, so that it suffices to consider subdirect products of $\{\mathfrak{A}_i \mid i \in I\}$.

Lemma 2.6 Suppose $\mathcal{C} \subseteq \mathcal{B} \subseteq \Pi\langle \mathfrak{A}_i \mid i \in I\rangle$ are both subdirect and $\mathfrak{T}$ is a topology on I containing the equalizer topology induced by $\mathcal{B}$. Then

$$\Gamma(\mathcal{C}, \mathfrak{T}) = \Gamma(\mathcal{B}, \mathfrak{T}).$$

Proof: Suppose $x \in \Gamma(C, \mathfrak{T})$ and $y \in B$. Since $\mathcal{C}$ is subdirect,

$$E(x, y) = \bigcup_{z \in C} E(x, z) \cap E(z, y) \in \mathfrak{T}.$$

Thus $x \in \Gamma(B, \mathfrak{T})$ and we have established that $\Gamma(\mathcal{C}, \mathfrak{T}) \subseteq \Gamma(\mathcal{B}, \mathfrak{T})$. Next suppose $x \in \Gamma(B, \mathfrak{T})$ and $y \in C$. Since $\mathcal{C} \subseteq \mathcal{B}$, $E(x, y) \in \mathfrak{T}$. Thus $x \in \Gamma(C, \mathfrak{T})$ and we have established that $\Gamma(\mathcal{B}, \mathfrak{T}) \subseteq \Gamma(\mathcal{C}, \mathfrak{T})$. ∎

Corollaries 2.2 and 2.5 and Lemma 2.6 show that for a fixed topology $\mathfrak{T}$ on I, $\Gamma(\cdot, \mathfrak{T})$ is a rather special closure operation on the set of all subdirect products of $\{\mathfrak{A}_i \mid i \in I\}$ inducing an equalizer topology contained in $\mathfrak{T}$. Accordingly we shall call $\Gamma(\mathcal{B}, \mathfrak{T})$ the $\mathfrak{T}$-global closure of $\mathcal{B}$. Of course, there may be no subdirect product of $\{\mathfrak{A}_i \mid i \in I\}$ inducing an equalizer topology contained in $\mathfrak{T}$. However, if there are such subdirect products then they are divided into disjoint cones (under the substructure relation), where on top of each cone sits a $\mathfrak{T}$-global subdirect product of $\{\mathfrak{A}_i \mid i \in I\}$, and underneath each top are those subdirect products whose $\mathfrak{T}$-global closure sits on top. In other words, if $\mathcal{B}$ is a subdirect product of $\{\mathfrak{A}_i \mid i \in I\}$ and $\mathfrak{T}$ is a topology on I containing the equalizer topology induced by $\mathcal{B}$ then $\mathcal{B}$ is $\mathfrak{T}$-global if and only if $\mathcal{B}$ is a maximal subdirect product of $\{\mathfrak{A}_i \mid i \in I\}$ inducing an equalizer topology included in $\mathfrak{T}$. Notice also that it does not make sense to speak of "the" $\mathfrak{T}$-global subdirect product of $\{\mathfrak{A}_i \mid i \in I\}$ because, first of all, there may be none, and then in case there is one there usually are many.

Lemma 2.7 Suppose $\mathcal{B} \subseteq \Pi\langle \mathfrak{A}_i \mid i \in I\rangle$ and $\mathfrak{T}_1$ and $\mathfrak{T}_2$ are topologies on I, where $\mathfrak{T}_1 \subseteq \mathfrak{T}_2$ and $\mathfrak{T}_1$ contains the equalizer topology induced by $\mathcal{B}$. Then $\Gamma(\mathcal{B}, \mathfrak{T}_1) \subseteq \Gamma(\mathcal{B}, \mathfrak{T}_2)$.

Proof: Suppose $x \in \Gamma(B, \mathfrak{T}_1)$. Then for every $y \in B$, $E(x, y) \in \mathfrak{T}_1 \subseteq \mathfrak{T}_2$ and therefore $x \in \Gamma(B, \mathfrak{T}_2)$. ∎

$\mathcal{B}$ is called a global substructure of $\Pi\langle \mathfrak{A}_i \mid i \in I\rangle$ if for some topology $\mathfrak{T}$ on I, $\mathcal{B}$ is a $\mathfrak{T}$-global substructure of $\Pi\langle \mathfrak{A}_i \mid i \in I\rangle$.

Corollary 2.8 Suppose $\mathcal{B} \subseteq \Pi\langle \mathfrak{A}_i \mid i \in I\rangle$ and $\mathfrak{T}$ is the equalizer topology on I induced by $\mathcal{B}$. Then $\mathcal{B}$ is global iff $\mathcal{B} = \Gamma(\mathcal{B}, \mathfrak{T})$.

Proof: Use Lemma 2.7. ∎

If $\mathcal{B}$ is global then the equalizer topology induced by $\mathcal{B}$ is the smallest topology with respect to which $\mathcal{B}$ is global. It may very well happen that $\mathcal{B}$ is global with respect to a topology which is larger than the equalizer topology induced by $\mathcal{B}$, and it appears that this fact may constitute valuable additional information. However, as we noticed before, if $\mathcal{B}$ is $\mathfrak{T}$-global then $\mathfrak{T}$ does not contain the equalizer topology induced by any subdirect product with the same factors which is larger than $\mathcal{B}$, so that in a sense $\mathfrak{T}$ cannot be much larger than the equalizer topology induced by $\mathcal{B}$.

If $\mathcal{B} \subseteq \Pi\langle \mathfrak{A}_i \mid i \in I\rangle$ and $\mathfrak{T}$ is the equalizer topology on I induced by $\mathcal{B}$ then we define

$$\Gamma(\mathcal{B}) = \Gamma(\mathcal{B}, \mathfrak{T}).$$

Then we obtain

(i) $\mathcal{B} \subseteq \Gamma(\mathcal{B}) \subseteq \Pi\langle \mathfrak{A}_i \mid i \in I\rangle$.

(ii) $\Gamma(\Gamma(\mathcal{B})) = \Gamma(\mathcal{B})$.

(iii) If $\mathcal{C} \subseteq \mathcal{B} \subseteq \Pi\langle \mathfrak{A}_i \mid i \in I\rangle$ are both subdirect then $\Gamma(\mathcal{C}) \subseteq \Gamma(\mathcal{B})$.

Indeed, (i) and (ii) follow from Corollaries 2.2 and 2.5. To verify (iii), let $\mathfrak{T}$ and $\mathfrak{T}'$ be the equalizer topologies induced by $\mathcal{C}$ and $\mathcal{B}$ respectively. Then by Lemmas 2.6 and 2.7,

$$\Gamma(\mathcal{C}) = \Gamma(\mathcal{C}, \mathfrak{T}) \subseteq \Gamma(\mathcal{C}, \mathfrak{T}') = \Gamma(\mathcal{B}, \mathfrak{T}') = \Gamma(\mathcal{B}).$$

Thus Γ is a closure operation on the subdirect products of $\{\mathfrak{A}_i \mid i \in I\}$. Accordingly we shall call $\Gamma(\mathcal{B})$ the global closure of $\mathcal{B}$. By Lemma 2.7, this is the smallest $\mathfrak{T}$-global closure of $\mathcal{B}$. The global subdirect products are exactly the Γ-closed subdirect products.

We shall illustrate the notion of globality by considering the three most natural topologies on an arbitrary set: the trivial topology, the discrete topology and the topology of cofinite sets.

Lemma 2.9 If $\mathcal{B} \subseteq \Pi\langle \mathfrak{A}_i \mid i \in I\rangle$ is subdirect then the following are equivalent:

(i) The equalizer topology induced by $\mathfrak{B}$ is trivial.

(ii) There exist $\mathfrak{A}$ and isomorphisms $h_i : \mathfrak{A} \to \mathfrak{A}_i$, for each $i \in I$, such that $B = \{h(a) \mid a \in A\}$, where the canonical embedding $h : \mathfrak{A} \to \Pi\langle \mathfrak{A}_i \mid i \in I\rangle$ is defined by

$$h(a)(i) = h_i(a) \quad \text{for all} \quad i \in I.$$

(iii) $\mathfrak{B}$ is $\{\emptyset, I\}$ -global.

Proof: Assume (i). Let φ be an atomic formula and let $x \in B^n$. If $[\![\varphi(x)]\!] \neq I$ then $[\![\varphi(x)]\!] = \emptyset$. It follows that for every $i \in I$, $\pi_i : \mathfrak{B} \to \mathfrak{A}_i$ is an isomorphism. Now pick fixed $k \in I$ and for each $i \in I$ define $h_i = \pi_i \circ \pi_k^{-1}$. Then $h_i : \mathfrak{A}_k \to \mathfrak{A}_i$ is an isomorphism and $B = \{h(a) \mid a \in A_k\}$. Indeed, if $a \in A_k$ then for each $i \in I$

$$h(a)(i) = \pi_i(\pi_k^{-1}(a)) = \pi_k^{-1}(a)(i)$$

and therefore $h(a) \in B$. Conversely, if $x \in B$ then $\pi_k(x) \in A_k$ and $h(\pi_k(x)) = x$. This establishes (ii).

Assume (ii). Let φ be an atomic formula and let $x \in A^n$. If $i \in [\![\varphi(h(x))]\!]$, then $\mathfrak{A}_i \models \varphi[\pi_i(h(x))]$. Thus $\mathfrak{A}_i \models \varphi[h_i(x)]$ and therefore $\mathfrak{A} \models \varphi[x]$. It follows that $[\![\varphi(h(x))]\!] = I$, and (i) is established.

Again assume (i). Suppose $x \in \Pi A_i$ and for all $y \in B$, $E(x, y) \in \{\emptyset, I\}$. Since $\mathfrak{B}$ is subdirect, there exists $y \in B$ such that $E(x, y) \neq \emptyset$ and therefore $x = y$. This establishes (iii), and (iii) implies (i) trivially. ∎

Lemma 2.9 says that the equalizer topology induced by $\mathfrak{B}$ is trivial just in case $\mathfrak{B}$ "essentially" is the diagonal substructure of a direct power. Indeed, if $\mathfrak{A} = \mathfrak{A}_i$ for all $i \in I$ then $\mathfrak{B}$ is a "twisted" diagonal substructure of $\mathfrak{A}^I$ and if, moreover, h_i is the identity mapping for each $i \in I$ then $\mathfrak{B}$ is the diagonal substructure of $\mathfrak{A}^I$.

Lemma 2.10 Suppose $\mathfrak{B} \subseteq \Pi\langle \mathfrak{A}_i \mid i \in I\rangle$ is subdirect. Then $\mathfrak{B}$ is global with respect to the discrete topology if and only if $\mathfrak{B} = \Pi\langle \mathfrak{A}_i \mid i \in I\rangle$.

Proof: Suppose $\mathfrak{B}$ is global with respect to the discrete topology. Then for every $x \in \Pi A_i$ and every $y \in B$, $E(x, y)$ is open. Thus $x \in B$. ∎

The smallest topology (trivial) yields the smallest subdirect representation (diagonal) and the largest topology (discrete) yields the largest subdirect representation (direct product). Notice that both representations must be considered extremely tight; the first one because it is isomorphic to its factors and the second one because it is a direct product. Between the two extreme topologies lies the topology of cofinite sets which yields the <u>direct sum representation</u>. In this case we shall only consider <u>algebras</u>.

Suppose for each $i \in I$, $\mathfrak{A}_i$ is an algebra and $\mathcal{E} \subseteq \Pi\langle \mathfrak{A}_i \mid i \in I\rangle$ is a trivial subalgebra determined by $e \in \Pi A_i$. Then we can form <u>the direct sum</u> $\Sigma_{\mathcal{E}}\langle \mathfrak{A}_i \mid i \in I\rangle$ of the family $\{\mathfrak{A}_i \mid i \in I\}$ <u>with respect to</u> $\mathcal{E}$, where

$$\Sigma_{\mathcal{E}} A_i = \{x \in \Pi A_i \mid D(x, e) \text{ is finite}\}.$$

$\mathfrak{B}$ is called <u>a direct sum of</u> $\{\mathfrak{A}_i \mid i \in I\}$ if $\mathfrak{B}$ is the direct sum of $\{\mathfrak{A}_i \mid i \in I\}$ with respect to some trivial subalgebra of $\Pi\langle \mathfrak{A}_i \mid i \in I\rangle$.

<u>Theorem</u> 2.11 If $\mathcal{E} \subseteq \Pi\langle \mathfrak{A}_i \mid i \in I\rangle$ is trivial and $\mathfrak{T}$ is the topology of cofinite subsets of I then $\Sigma_{\mathcal{E}}\langle \mathfrak{A}_i \mid i \in I\rangle$ is a $\mathfrak{T}$-global subdirect product of $\{\mathfrak{A}_i \mid i \in I\}$.

<u>Proof</u>: If $x, y \in \Sigma_{\mathcal{E}} A_i$ then $E(x, e) \cap E(x, e) \subseteq E(x, y) \in \mathfrak{T}$. If $x \in \Pi A_i$ and $E(x, y) \in \mathfrak{T}$ for all $y \in \Sigma_{\mathcal{E}} A_i$, then $E(x, e) \in \mathfrak{T}$ and therefore $x \in \Sigma_{\mathcal{E}} A_i$. ∎

More generally, an algebra $\mathfrak{B} \subseteq \Pi\langle \mathfrak{A}_i \mid i \in I\rangle$ is called <u>a weak direct product of</u> $\{\mathfrak{A}_i \mid i \in I\}$ (Grätzer [17]) if

(i) for all $x, y \in B$, $D(x, y)$ is finite;

(ii) for all $x \in B$ and $y \in \Pi A_i$, if $D(x, y)$ is finite then $y \in B$.

In case the similarity type is finitely based weak direct products hardly amount to a significant generalization of direct sums.

<u>Lemma</u> 2.12 If Op is finite then an algebra $\mathfrak{B}$ is a weak direct product of $\{\mathfrak{A}_i \mid i \in I\}$ if and only if there exist finite $J \subseteq I$ and trivial $\mathcal{E} \subseteq \Pi\langle \mathfrak{A}_i \mid i \in I - J\rangle$ such that

$$\mathfrak{B} = \Pi\langle \mathfrak{A}_j \mid j \in J\rangle \times \Sigma_{\mathcal{E}}\langle \mathfrak{A}_i \mid i \in I - J\rangle.$$

<u>Proof</u>: It is obvious that

$$\mathfrak{B} = \Pi\langle \mathfrak{A}_j \mid j \in J\rangle \times \Sigma_{\mathcal{E}}\langle \mathfrak{A}_i \mid i \in I - J\rangle$$

is a weak direct product of $\{\mathfrak{A}_i \mid i \in I\}$. Conversely, suppose $\mathfrak{B}$ is a weak direct product of $\{\mathfrak{A}_i \mid i \in I\}$ and choose any $b \in B$. Let

$$J = \bigcup\{D(f^{\mathfrak{B}}(b, b, \ldots, b), b) \mid f \in Op\}.$$

Since Op is finite and $\mathfrak{B}$ is a weak direct product, J is finite. Now $\pi_{I-J}(b)$ determines a trivial subalgebra $\mathfrak{E} \subseteq \Pi\langle \mathfrak{A}_i \mid i \in I - J\rangle$. If

$$\mathfrak{C} = \Pi\langle \mathfrak{A}_j \mid j \in J\rangle \times \Sigma_{\mathfrak{E}}\langle \mathfrak{A}_i \mid i \in I - J\rangle$$

then $\mathfrak{B} \subseteq \mathfrak{C}$ by (i) of the definition of weak direct products, and $\mathfrak{C} \subseteq \mathfrak{B}$ by (ii). ∎

Corollary 2.13 If Op is finite and each $\mathfrak{A}_i$ contains a trivial subalgebra then every weak direct product of $\{\mathfrak{A}_i \mid i \in I\}$ is a direct sum of $\{\mathfrak{A}_i \mid i \in I\}$.

Proof: Let $\mathfrak{B} = \Pi\langle \mathfrak{A}_j \mid j \in J\rangle \times \Sigma_{\mathfrak{E}}\langle \mathfrak{A}_i \mid i \in I - J\rangle$, where $J \subseteq I$ is finite and $\mathfrak{E} \subseteq \Pi\langle \mathfrak{A}_i \mid i \in I - J\rangle$ is trivial. If $\mathfrak{E}' \subseteq \Pi\langle \mathfrak{A}_j \mid j \in J\rangle$ is trivial then

$$\mathfrak{B} = \Sigma_{\mathfrak{E}' \times \mathfrak{E}}\langle \mathfrak{A}_i \mid i \in I\rangle. \blacksquare$$

Theorem 2.14 Suppose an algebra $\mathfrak{B} \subseteq \Pi\langle \mathfrak{A}_i \mid i \in I\rangle$ is subdirect, and let $\mathfrak{T}$ be the topology of cofinite subsets of I. Then $\mathfrak{B}$ is $\mathfrak{T}$-global if and only if $\mathfrak{B}$ is a weak direct product of $\{\mathfrak{A}_i \mid i \in I\}$.

Proof: Suppose $\mathfrak{B} = \Gamma(\mathfrak{B}, \mathfrak{T})$ and let $x \in B$ and $y \in \Pi A_i$ where $E(x, y) \in \mathfrak{T}$. Consider any $z \in B$. Then

$$E(z, x) \cap E(x, y) \subseteq E(z, y) \in \mathfrak{T}$$

and therefore $y \in B$. The converse follows directly from the definition. ∎

Lemma 2.15 Suppose an algebra $\mathfrak{B} \subseteq \Pi\langle \mathfrak{A}_i \mid i \in I\rangle$ is subdirect and $\mathfrak{B}$ is $\mathfrak{T}$-global, where $\mathfrak{T}$ is the topology of cofinite subsets of I. Then $\mathfrak{T}$ is the equalizer topology induced by $\mathfrak{B}$ if and only if every $\mathfrak{A}_i$ is non-trivial.

Proof: Suppose $\mathfrak{T}$ is the equalizer topology induced by $\mathfrak{B}$ and let $i \in I$. Since $I - \{i\}$ is open, there exist $r < \omega$ and $x_p, y_p \in B$ for each $p < r$ such that

$$\bigcap_{p<r} E(x_p, y_p) \subseteq I - \{i\}.$$

For some $p < r$, $i \notin E(x_p, y_p)$ and therefore $x_p(i) \neq y_p(i)$. Conversely, suppose every $\mathfrak{A}_i$ is non-trivial and let $J \in \mathfrak{S}$. Pick $b \in B$ and define

$$x(i) = \begin{cases} b(i) & \text{if } i \in J \\ c(i) \neq b(i) & \text{if } i \notin J \end{cases}$$

Then $J = E(x, b)$ and by Theorem 2.14, $x \in B$. ∎

The special feature of weak direct products of non-trivial algebras is that the equalizer topology is actually a <u>filter</u>. Generalizing this idea one quickly realizes that it suffices to consider filters containing all cofinite sets. This way one obtains global subdirect products which are larger than weak direct products (see Lemma 2.7), as was done by Hashimoto [18] (see also Grätzer [17]). Let $\mathfrak{J}$ be an ideal in the power set of I containing all finite sets. Then an algebra $\mathfrak{B} \subseteq \Pi\langle \mathfrak{A}_i \mid i \in I\rangle$ is called a <u>$\mathfrak{J}$-restricted direct product</u> of $\{\mathfrak{A}_i \mid i \in I\}$ if

(i) for all $x, y \in B$, $D(x, y) \in \mathfrak{J}$;

(ii) for all $x \in B$ and $y \in \Pi A_i$, if $D(x, y) \in \mathfrak{J}$ then $y \in B$.

Let

$$\breve{\mathfrak{J}} = \{I - F \mid F \in \mathfrak{J}\}.$$

Then $\breve{\mathfrak{J}}$ is a topology (filter) on I containing the topology of cofinite sets.

<u>Theorem</u> 2.16 Suppose an algebra $\mathfrak{B} \subseteq \Pi\langle \mathfrak{A}_i \mid i \in I\rangle$ is subdirect, and let $\mathfrak{J}$ be an ideal in the power set of I containing all finite sets. Then $\mathfrak{B}$ is $\breve{\mathfrak{J}}$-global if and only if $\mathfrak{B}$ is an $\mathfrak{J}$-restricted direct product of $\{\mathfrak{A}_i \mid i \in I\}$. ∎

Comer [9] first noticed the connection between the direct sum construction and the sheaf construction.

Next we shall illustrate the notion of global closure. Diagonal substructures of direct powers are global, but with respect to a very small topology (Lemma 2.9). Thus we can take global closure with respect to a larger topology. This leads to another familiar construction from universal algebra.

Let $\mathfrak{A}$ be a non-trivial structure, give A the discrete topology, and let $\mathfrak{S}$ be a topology on I. For each $a \in A$, let $\bar{a} = I \times \{a\}$ be the <u>diagonal element</u> of A^I corresponding to a, and let $\bar{\mathfrak{A}}$ be the <u>diagonal substructure</u> of $\mathfrak{A}^I$ determined by the diagonal $\bar{A} = \{\bar{a} \mid a \in A\}$. Finally, let $C_{\mathfrak{S}}(I, A)$ be the set of continuous $x \in A^I$. From Lemma 2.9 recall that the equalizer topology induced by $\overline{\mathfrak{A}}$ is trivial and therefore is contained in $\mathfrak{T}$.

Lemma 2.17 $C_{\mathfrak{T}}(I, A) = \Gamma(\overline{A}, \mathfrak{T})$.

Proof: Suppose $x \in C_{\mathfrak{T}}(I, A)$ and $a \in A$. Then

$$E(x, \overline{a}) = \{i \in I \mid x(i) = a\} \in \mathfrak{T}.$$

Thus $x \in \Gamma(\overline{A}, \mathfrak{T})$. The proof of the reverse inclusion is just as trivial. ∎

As is well-known (and follows from Lemma 2.17), $C_{\mathfrak{T}}(I, A)$ determines a subalgebra of $\mathfrak{A}^I$ which is denoted by $C_{\mathfrak{T}}(I, \mathfrak{A})$.

Corollary 2.18 $C_{\mathfrak{T}}(I, \mathfrak{A}) = \Gamma(\overline{\mathfrak{A}}, \mathfrak{T})$. ∎

Theorem 2.19 Let $\mathfrak{T}$ be a topology on I. If $\mathfrak{B} \subseteq \mathfrak{A}^I$ is $\mathfrak{T}$-global then $\overline{\mathfrak{A}} \subseteq \mathfrak{B}$ if and only if $\mathfrak{B} = \Gamma(\overline{\mathfrak{A}}, \mathfrak{T})$.

Proof: Suppose $\overline{\mathfrak{A}} \subseteq \mathfrak{B}$. By Lemma 2.6, $\mathfrak{B} = \Gamma(\mathfrak{B}, \mathfrak{T}) \subseteq \Gamma(\overline{\mathfrak{A}}, \mathfrak{T})$. Moreover, if $x \in \Gamma(\overline{A}, \mathfrak{T})$ and $y \in B$ then

$$E(x, y) = \bigcup_{a \in A} E(x, \overline{a}) \cap E(y, \overline{a}) \in \mathfrak{T}$$

and therefore $x \in B$. Thus $\Gamma(\overline{\mathfrak{A}}, \mathfrak{T}) \subseteq \mathfrak{B}$. This proves one direction of the assertion and the other is trivial. ∎

If $\mathfrak{F}$ is the field of clopen sets of I then $C_{\mathfrak{F}}(I, A)$ is exactly the set of all $x \in A^I$ such that for all $a \in A$, $x^{-1}(a) \in \mathfrak{F}$, and $C_{\mathfrak{F}}(I, \mathfrak{A})$ is usually called a Boolean power of $\mathfrak{A}$ (As it stands, it is customary in the literature to require that I is a Boolean space, but we obviously can take for $\mathfrak{F}$ the field of all clopen sets of any topological space).

Theorem 2.20 Let $\mathfrak{T}$ be a topology on I and let $\mathfrak{B} \subseteq C_{\mathfrak{T}}(I, \mathfrak{A})$. Then $\mathfrak{B}$ is $\mathfrak{T}$-global if and only if

$$B = C_{\mathfrak{T}}(I, A) \cap \Pi\langle \pi_i(B) \mid i \in I\rangle.$$

Proof: Assume $\mathfrak{B}$ is $\mathfrak{T}$-global. Let $x \in C_{\mathfrak{T}}(I, A) \cap \Pi\pi_i(B)$ and consider $y \in B$. Then

$$E(x, y) = \bigcup_{a \in A} E(x, \overline{a}) \cap E(y, \overline{a}) \in \mathfrak{T},$$

and therefore $x \in B$. Conversely, assume the right-hand side of the assertion.

Suppose $x \in \Pi\pi_i(B)$ and for every $y \in B$, $E(x, y) \in \mathfrak{F}$. Consider any $a \in A$. For each $i \in E(x, \bar{a})$ there exists $y_i \in B$ such that $i \in E(x, \bar{a}) \cap E(x, y_i)$. Thus

$$i \in E(y_i, \bar{a}) \cap E(x, y_i) \subseteq E(x, \bar{a})$$

and therefore

$$E(x, \bar{a}) = \bigcup\{E(y_i, \bar{a}) \cap E(x, y_i) \mid i \in E(x, \bar{a})\} \in \mathfrak{F}.$$

This establishes that $x \in C_{\mathfrak{F}}(I, A)$. It follows from the hypothesis that $x \in B$, and $\mathfrak{B}$ is $\mathfrak{F}$-global. ∎

$\mathfrak{F}$-global substructures of $C_{\mathfrak{F}}(I, \mathfrak{A})$ are closely related to <u>sub-Boolean</u> powers of $\mathfrak{A}$ (Burris [6]) or <u>filtered Boolean powers</u> of $\mathfrak{A}$ (Burris and Werner [5]). The connection is easy to see because for finite $\mathfrak{A}$ there exists $r < \omega$ such that

$$\{\pi_i(\mathfrak{B}) \mid i \in I\} = \{\mathfrak{A}_p \mid p < r\}.$$

Now for each $p < r$,

$$\begin{aligned} J_p &= \{i \in I \mid \pi_i(\mathfrak{B}) \subseteq \mathfrak{A}_p\} \\ &= \bigcap_{b \in B} \bigcup_{a \in A_p} E(b, \bar{a}) \end{aligned}$$

is closed and

$$B = \{x \in C_{\mathfrak{F}}(I, A) \mid \text{for all } p < r,\ x(J_p) \subseteq A_p\}.$$

Conversely, if for each $p < r$, $\mathfrak{A}_p \subseteq \mathfrak{A}$ and $J_p \subseteq I$ is closed, and if

$$B = \{x \in C_{\mathfrak{F}}(I, A) \mid \text{for all } p < r,\ x(J_p) \subseteq A_p\},$$

then $B = C_{\mathfrak{F}}(I, A) \cap \Pi\langle\pi_i(B) \mid i \in I\rangle$. Thus for <u>finite</u> $\mathfrak{A}$ Theorem 2.20 exactly yields the filtered Boolean powers of $\mathfrak{A}$. (It is again not necessary to require that I is a Boolean space, as is customary in the literature.)

We shall continue our discussion of direct sums and Boolean powers in Section 5. Now we shall turn to the main topic of this section. To relate global subdirect products to the structures of global sections of sheaves we have to review (hopefully for the last time) the basic properties of this elaborate construction. A (discrete) <u>sheaf of structures</u> is a triple $\mathcal{S} = \langle S, \eta, I\rangle$, where

(i) S and I are topological spaces and $\eta: S \to I$ is a surjective local homeomorphism, that is for each $a \in S$ there exists an open neighborhood G of a such that $\eta(G)$ is open and the restriction of η to G is a homeomorphism;

(ii) for each $i \in I$, $\eta^{-1}(i)$ is the universe S_i of a structure $\mathfrak{A}_i$;

(iii) for each $(n + 1)$-ary operation symbol f, the mapping

$$\bigcup_{i\in I} f^{\mathfrak{A}_i} : \bigcup_{i\in I} S_i^{n+1} \to S$$

is continuous, where $\bigcup_{i\in I} S_i^{n+1}$ inherits the product topology from S^{n+1};

(iv) for each individual constant c, the mapping

$$c^{\Pi \mathfrak{A}_i} : I \to S$$

is continuous;

(v) for each $(n + 1)$-ary relation symbol R, $\bigcup_{i\in I} R^{\mathfrak{A}_i}$ is an open subset of $\bigcup_{i\in I} S_i^{n+1}$.

The structures $\mathfrak{A}_i$ are called the <u>stalks</u> of the sheaf, the space S is called the <u>stalk space</u> (or <u>total space</u>), and the space I is called the <u>index space</u> (or <u>base space</u>). A <u>global section</u> of $\mathcal{S}$ is a continuous function $x:I \to S$ such that for every $i \in I$, $\eta(x(i)) = i$. If the set of global sections of $\mathcal{S}$ is non-empty then it determines a substructure

$$\Gamma(\mathcal{S}) \subseteq \Pi\langle \mathfrak{A}_i \mid i \in I\rangle,$$

which is called the <u>structure of global sections</u> of $\mathcal{S}$. The next observations are well-known and easy to prove.

<u>Lemma</u> 2.21 If $\mathcal{S} = \langle S, \eta, I\rangle$ is a sheaf of structures then η is an open mapping. ∎

<u>Lemma</u> 2.22 If $\mathcal{S} = \langle S, \eta, I\rangle$ is a sheaf of structures and $x \in \Pi\langle S_i \mid i \in I\rangle$ then the following are equivalent:

(i) $x \in \Gamma(\mathcal{S})$.

(ii) For each $i \in I$ there exists open $G_i \subseteq S$ such that $x(i) \in G_i \subseteq x(I)$, $\eta(G_i)$ is open and the restriction of η to G_i is a homeomorphism.

(iii) x is a local homeomorphism.

(iv) x is an open mapping. ∎

<u>Lemma</u> 2.23 For every atomic formula φ and every $x \in \Gamma(\mathcal{S})^n$, $[\![\varphi(x)]\!]$ is open. ∎

Lemma 2.24 For every $i \in I$, the relative topology on $\pi_i(\Gamma(\mathcal{S}))$ is discrete. ∎

Lemma 2.24 explains the terminology "discrete" sheaf construction. In so-called "topological" sheaf constructions the stalks are endowed with some natural topology which enters into the definition of the sheaf. This is a still more involved construction which will not be touched upon in this paper.

If $\mathcal{B} \subseteq \Pi\langle \mathfrak{A}_i \mid i \in I\rangle$ then we say that $\mathcal{B}$ is disjointed if $\pi_i(B) \cap \pi_j(B) = \emptyset$ whenever $i \neq j$. Of course, if $\mathcal{S} = \langle S, \eta, I\rangle$ is a sheaf of structures then $\Gamma(\mathcal{S})$ is a disjointed substructure of $\Pi\langle \mathcal{S}_i \mid i \in I\rangle$.

Lemma 2.25 Suppose $\mathcal{B} \subseteq \Pi\langle \mathfrak{A}_i \mid i \in I\rangle$ is disjointed. Then for any $U \subseteq I$ and any $x, y \in \Pi\pi_i(B)$,

$$x^{-1}(y(U)) = U \cap E(x, y).$$

Proof: $x(i) \in y(U)$ if and only if for some $j \in U$, $x(i) = y(j)$ and, since $\pi_i(B) \cap \pi_j(B) = \emptyset$ for $i \neq j$, this is the case if and only if $i \in U$ and $x(i) = y(i)$. ∎

Lemma 2.26 Let $\mathcal{S} = \langle S, \eta, I\rangle$ be a sheaf of structures. If $\mathfrak{T}$ is the topology on I then $\Gamma(\mathcal{S})$ is a disjointed $\mathfrak{T}$-global substructure of $\Pi\langle \mathcal{S}_i \mid i \in I\rangle$.

Proof: Suppose $x \in \Pi\langle\pi_i(\Gamma(\mathcal{S})) \mid i \in I\rangle$. Define

$$S^o = \bigcup\{y(I) \mid y \in \Gamma(\mathcal{S})\}.$$

By Lemma 2.21, S^o is open. Since $x(I) \subseteq S^o$, $x \in \Gamma(\mathcal{S})$ if and only if for all open $G \subseteq S^o$, $x^{-1}(G) \in \mathfrak{T}$. If $G \subseteq S^o$ then

$$G = \bigcup\{y(y^{-1}(G)) \mid y \in \Gamma(\mathcal{S})\}.$$

Moreover, if $G \subseteq S^o$ is open and $y \in \Gamma(\mathcal{S})$ then $y^{-1}(G) \in \mathfrak{T}$ and therefore, by Lemma 2.22, $y(y^{-1}(G))$ is open. Thus $x \in \Gamma(\mathcal{S})$ if and only if for all $U \in \mathfrak{T}$ and all $y \in \Gamma(\mathcal{S})$, $x^{-1}(y(U)) \in \mathfrak{T}$. By Lemma 2.25, $x \in \Gamma(\mathcal{S})$ if and only if for all $U \in \mathfrak{T}$ and all $y \in \Gamma(\mathcal{S})$, $U \cap E(x, y) \in \mathfrak{T}$. Thus $x \in \Gamma(\mathcal{S})$ if and only if for all $y \in \Gamma(\mathcal{S})$, $E(x, y) \in \mathfrak{T}$. The assertion follows from Corollary 2.3 and Lemma 2.23. ∎

Now let $\mathcal{B} \subseteq \Pi\langle \mathfrak{A}_i \mid i \in I\rangle$ be disjointed and let $\mathfrak{T}$ be a topology on I containing the equalizer topology induced by $\mathcal{B}$. We carry out the standard sheaf construction based on $\mathcal{B}$ and $\mathfrak{T}$ as follows: Define

$$S = \bigcup_{i \in I} \pi_i(B)$$

and give S the topology with basis

$$\{x(U) \mid x \in B,\ U \in \mathcal{T}\}.$$

Define

$$\eta(a) = i, \quad \text{where} \quad a \in \pi_i(B).$$

Then $\mathcal{S} = \langle S, \eta, I \rangle$ is a sheaf of structures. The standard sheaf construction has been carried out in one form or another by almost everybody who has worked in the field, and apparently has become part of the folklore of sheaf representation theory. The first who has carried out this construction in the setting of universal algebra appears to be Davey [13].

<u>Lemma</u> 2.27 $\Gamma(\mathcal{S}) = \Gamma(\mathcal{B}, \mathcal{T})$.

<u>Proof</u>: Suppose $x \in \Gamma(\mathcal{S})$ and consider $y \in B$. Since $B \subseteq \Gamma(\mathcal{S})$, by Lemma 2.23, $E(x, y) \in \mathcal{T}$ and therefore $x \in \Gamma(B, \mathcal{T})$. Conversely, suppose $x \in \Gamma(B, \mathcal{T})$ and let $y \in B$, $U \in \mathcal{T}$. By Lemma 2.25

$$x^{-1}(y(U)) = U \cap E(x, y) \in \mathcal{T}$$

and therefore $x \in \Gamma(\mathcal{S})$. ∎

Collecting our observations we obtain the main result of this section.

<u>Theorem</u> 2.28 Let $\mathcal{B} \subseteq \Pi\langle \mathfrak{A}_i \mid i \in I \rangle$ be subdirect, and let $\mathcal{T}$ be the equalizer topology on I induced by $\mathcal{B}$. Then the following are equivalent:

(i) $\mathcal{B}$ is a global subdirect product.

(ii) If $x \in \Pi A_i$ and $E(x, y) \in \mathcal{T}$ for all $y \in B$, then $x \in B$.

If, moreover, $\mathcal{B}$ is disjointed then (i) and (ii) are equivalent to

(iii) $\mathcal{B} = \Gamma(\mathcal{S})$ for some sheaf $\mathcal{S}$.

<u>Proof</u>: Use Corollary 2.8 and Lemma 2.26 and 2.27. ∎

It is clear that global closure is a generalization of the standard sheaf construction which releaves this construction of the requirement that the factors be disjointed. By passing from structures of global sections of sheaves of structures to global substructures of direct products we have enlarged the class of objects of the category. We shall now show that in a very strict sense we have not added any new "isomorphism types". Let $\mathcal{S} = \langle S, \eta, I \rangle$ and $\mathcal{S}' = \langle S', \eta', I' \rangle$ be sheaves of structures. $\mathcal{S}$ and $\mathcal{S}'$ are called <u>iso-</u>

morphic if there are a homeomorphism f from I onto I' and a homeomorphism g from S onto S' such that

(i) $f \circ \eta = \eta' \circ g$;

(ii) for each $i \in I$, the restriction of g to S_i is an isomorphism from $\mathcal{Y}_i$ onto $\mathcal{Y}_i'$.

Correspondingly, Let $\mathcal{B} \subseteq \Pi\langle \mathfrak{A}_i \mid i \in I\rangle$, $\mathcal{B}' \subseteq \Pi\langle \mathfrak{A}_j' \mid j \in I'\rangle$ and let $\mathfrak{T}$ and $\mathfrak{T}'$ be topologies on I and I' respectively. We define

$$\mathcal{B}\ {}_{\mathfrak{T}}\cong_{\mathfrak{T}'}\ \mathcal{B}'$$

if $\mathcal{B}$ and $\mathcal{B}'$ are $\mathfrak{T}$- and $\mathfrak{T}'$-global respectively and there exists a homeomorphism f from I onto I', and for each $i \in I$ there exists an isomorphism h_i from $\pi_i(\mathcal{B})$ onto $\pi_{f(i)}(\mathcal{B}')$ such that the canonical mapping h defined for $x \in B$ by

$$h(x)(f(i)) = h_i(\pi_i(x))$$

is an isomorphism from $\mathcal{B}$ onto $\mathcal{B}'$.

Theorem 2.29 Suppose $\mathcal{S} = \langle S, \eta, I\rangle$ and $\mathcal{S}' = \langle S', \eta', I'\rangle$ are sheaves of structures, where $\Gamma(\mathcal{S})$ and $\Gamma(\mathcal{S}')$ are subdirect products of the stalks respectively, and let $\mathfrak{T}$ and $\mathfrak{T}'$ be the topologies on I and I' respectively. Then $\mathcal{S}$ is isomorphic to $\mathcal{S}'$ if and only if $\Gamma(\mathcal{S})\ {}_{\mathfrak{T}}\cong_{\mathfrak{T}'}\ \Gamma(\mathcal{S}')$.

Proof: Assume $\mathcal{S}$ is isomorphic to $\mathcal{S}'$. Let f be the homeomorphism from I onto I' and let g be the homeomorphism from S onto S'. For each $i \in I$, let h_i be the restriction of g to S_i. Then the canonical mapping h maps $\Gamma(\mathcal{S})$ onto $\Gamma(\mathcal{S}')$.

Conversely, assume $\Gamma(\mathcal{S})\ {}_{\mathfrak{T}}\cong_{\mathfrak{T}'}\ \Gamma(\mathcal{S}')$. Let f be the homeomorphism from I onto I', and for each $i \in I$, let h_i be the isomorphism from $\mathcal{Y}_i$ onto $\mathcal{Y}'_{f(i)}$. Next define $g : S \to S'$ by

$$g(\pi_i(x)) = h_i(\pi_i(x)).$$

It is now tedious but easy to check that g is the desired homeomorphism from S onto S' to show that $\mathcal{S}$ and $\mathcal{S}'$ are isomorphic. ∎

Due to the special set theoretic construction of sheaves, structures of global sections are disjointed subdirect products. Arbitrary global subdirect products of course are not always disjointed. However, to an arbitrary direct product $\Pi\langle \mathfrak{A}_i \mid i \in I\rangle$ we can apply the device of disjointing the factors.

We shall give a quick review of this tedious and well-known process. For each $i \in I$ define $\mathfrak{A}_i'$ with universe $\{i\} \times A_i$ such that the mapping $h_i(a) = \langle i, a\rangle$ becomes an isomorphism from $\mathfrak{A}_i$ onto $\mathfrak{A}_i'$. Then the canonical mapping h defined for $x \in \Pi A_i$ by

$$h(x)(i) = h_i(x(i))$$

is an isomorphism from $\Pi\mathfrak{A}_i$ onto $\Pi\mathfrak{A}_i'$, and for each $x \in (\pi A_i)^\omega$ and each formula φ

$$[\![\varphi(x)]\!] = [\![\varphi(h(x))]\!] .$$

Hence $\mathcal{B} \subseteq \Pi\mathfrak{A}_i$ and $h(\mathcal{B}) \subseteq \Pi\mathfrak{A}_i'$ induce the same equalizer topology on I, and if $\mathfrak{T}$ is a topology on I containing the equalizer topology induced by $\mathcal{B}$ then

$$x \in \Gamma(B, \mathfrak{T}) \text{ iff } h(x) \in \Gamma(h(B), \mathfrak{T}).$$

Thus we obtain:

Lemma 2.30 $\mathcal{B} \subseteq \Pi\langle\mathfrak{A}_i \mid i \in I\rangle$ is $\mathfrak{T}$-global if and only if $h(\mathcal{B}) \subseteq \Pi\langle\mathfrak{A}_i' \mid i \in I\rangle$ is $\mathfrak{T}$-global, and in this case $\mathcal{B}\ {}_{\mathfrak{T}}\!\cong_{\mathfrak{T}'} h(\mathcal{B})$. ∎

Corollary 2.31 The structures of global sections of sheaves of structures are (up to isomorphism over the index topologies) exactly the global substructures of direct products. ∎

Passing from structures of global sections of sheaves of structures to global substructures of direct products is very natural and leads to considerable simplifications in algebraic constructions that are related to the sheaf construction. It is obvious that the disjointness of the stalks plays no essential role at all. To the contrary, disjointing the stalks is one of the more tedious and trivial aspects of the sheaf construction. Moreover, since all structures of global sections "essentially" can be obtained from the standard sheaf construction, it is plausible that the stalk topology plays no essential role either. In our setting we completely dispense with the stalk topology, and the central role of the equalizer topology becomes clearly visible.

Since some important properties of the sheaf construction are usually expressed in terms of the stalk topology, we shall paraphrase these properties eliminating reference to the stalk topology. This transcription usually simplifies the terminology considerably. For a first example the reader should return to Theorem 2.29 where we show that ${}_{\mathfrak{T}}\!\cong_{\mathfrak{T}'}$ is the correct analogue of <u>sheaf isomorphism</u>.

Next we shall discuss Hausdorff sheaves. A sheaf of structures $\mathcal{S} = \langle S, \eta, I\rangle$ is called <u>Hausdorff</u> if the stalk space S is Hausdorff.

<u>Lemma</u> 2.32 Suppose $\mathcal{S} = \langle S, \eta, I\rangle$ is a sheaf of structures and let

$$S^o = \bigcup\{x(I) \mid x \in \Gamma(\mathcal{S})\}.$$

Then S^o is a Hausdorff space (with the relative topology) if and only if I is Hausdorff and for every $x, y \in \Gamma(\mathcal{S})$, $E(x, y)$ is clopen.

<u>Proof</u>: Suppose S^o is Hausdorff and let $i, j \in I$, where $i \neq j$. Consider any $x \in \Gamma(\mathcal{S})$. Since $S_i \cap S_j = \emptyset$, $x(i) \neq x(j)$. Thus there exist disjoint open neighborhoods G and H of $x(i)$ and $x(j)$ respectively. It follows that $x^{-1}(G)$ and $x^{-1}(H)$ are disjoint open neighborhoods of i and j respectively. Moreover, it is well-known that $E(x, y)$ is closed for all $x, y \in \Gamma(\mathcal{S})$ in case S^o is Hausdorff. By Lemma 2.23, $E(x, y)$ is clopen for all $x, y \in \Gamma(\mathcal{S})$.

Conversely, assume the other side of the assertion and let $a, b \in S^o$, where $a \neq b$. We distinguish two cases. <u>Case</u> 1: $a \in S_i$ and $b \in S_j$, where $i \neq j$. Let U and V be disjoint open neighborhoods of i and j respectively and choose $x, y \in \Gamma(\mathcal{S})$ such that $x(i) = a$ and $y(j) = b$. Then, by Lemma 2.22 and the disjointness of the stalks, $x(U)$ and $x(V)$ are disjoint open neighborhoods of a and b respectively. <u>Case</u> 2: $a, b \in S_i$. Again choose $x, y \in \Gamma(\mathcal{S})$ such that $x(i) = a$ and $y(i) = b$. Then $i \in D(x, y)$, where $D(x, y)$ is open. Thus $x(D(x, y))$ and $y(D(x, y))$ are disjoint open neighborhoods of a and b respectively. ∎

<u>Corollary</u> 2.33 Let $\mathcal{S} = \langle S, \eta, I\rangle$ be a sheaf of structures.

(i) If $\mathcal{S}$ is Hausdorff then the index space I is Hausdorff and for every $x, y \in \Gamma(\mathcal{S})$, $E(x, y)$ is clopen.

(ii) If $\Gamma(\mathcal{S})$ is a subdirect product of the stalks, then $\mathcal{S}$ is Hausdorff if and only if I is Hausdorff and for every $x, y \in \Gamma(\mathcal{S})$, $E(x\ y)$ is clopen. ∎

Hausdorff sheaves play an important role in sheaf representation. We shall see later in this section (Theorem 2.39) that the index space of a global subdirect product can always be made Hausdorff in case all equalizers are clopen. <u>Thus "Hausdorff sheaf" essentially means equalizers are clopen</u>.

We proceed to discuss subsheaves. Let $\mathcal{S} = \langle S, \eta, I\rangle$ and $\mathcal{S}' = \langle S', \eta', I'\rangle$ be sheaves of structures. $\mathcal{S}'$ is called a <u>subsheaf</u> of $\mathcal{S}$ if S' is an open subset of S, $I' = I$, η' is the restriction of η to

S' and for every $i \in I$, $\mathcal{X}_i' \subseteq \mathcal{X}_i$.

Correspondingly, let $\mathcal{B}, \mathcal{C} \subseteq \Pi\langle \mathfrak{A}_i \mid i \in I\rangle$ and let $\mathcal{T}$ be a topology on I. We define

$$\mathcal{B} \subseteq_{\mathcal{T}} \mathcal{C}$$

if $\mathcal{B}$ and $\mathcal{C}$ are both $\mathcal{T}$-global substructures of $\Pi\langle \mathfrak{A}_i \mid i \in I\rangle$ and $\mathcal{B} \subseteq \mathcal{C}$.

<u>Theorem</u> 2.34 Suppose $\mathcal{S} = \langle S, \eta, I\rangle$ and $\mathcal{S}' = \langle S', \eta', I\rangle$ are sheaves of structures, where $\Gamma(\mathcal{S})$ and $\Gamma(\mathcal{S}')$ are subdirect products of the stalks respectively, and let $\mathcal{T}$ be the topology on I. Then $\mathcal{S}'$ is a subsheaf of $\mathcal{S}$ if and only if $\Gamma(\mathcal{S}') \subseteq_{\mathcal{T}} \Gamma(\mathcal{S})$.

<u>Proof</u>: Assume $\mathcal{S}'$ is a subsheaf of $\mathcal{S}$. Clearly $\Gamma(\mathcal{S}') \subseteq \Gamma(\mathcal{S})$ and by Lemma 2.26, $\Gamma(\mathcal{S})$ is a $\mathcal{T}$-global substructure of $\Pi\langle \mathcal{X}_i \mid i \in I\rangle$. Let $x \in \Pi\langle \pi_i(\Gamma(\mathcal{S}')) \mid i \in I\rangle$, where for every $y \in \Gamma(\mathcal{S}')$, $E(x, y) \in \mathcal{T}$. Consider $z \in \Gamma(\mathcal{S})$. Then

$$E(x, z) = \bigcup\{E(x, y) \cap E(y, z) \mid y \in \Gamma(\mathcal{S}')\} \in \mathcal{T}.$$

By Corollary 2.3, $x \in \Gamma(\mathcal{S})$ and it follows from the assumption that $x \in \Gamma(\mathcal{S}')$. This establishes that $\Gamma(\mathcal{S}')$ is a $\mathcal{T}$-global substructure of $\Pi\langle \mathcal{X}_i \mid i \in I\rangle$.

Conversely, assume $\Gamma(\mathcal{S}') \subseteq_{\mathcal{T}} \Gamma(\mathcal{S})$. By Lemma 2.22, S' is an open subset of S and it follows that $\mathcal{S}'$ is a subsheaf of $\mathcal{S}$. ∎

Theorem 2.34 establishes that $\subseteq_{\mathcal{T}}$ is the correct analogue of subsheaf. To give an example, let $\mathcal{S} = \langle S, \eta, I\rangle$ be a sheaf of structures. Then $\Gamma(\mathcal{S})$ is not in general a subdirect product of the stalks $\mathcal{X}_i$. However, by Lemma 2.22,

$$S^o = \bigcup\{x(I) \mid x \in \Gamma(\mathcal{S})\}$$

is an open subset of S and $\mathcal{S}^o = \langle S^o, \eta^o, I\rangle$ is a subsheaf of $\mathcal{S}$, where η^o is the restriction of η to S^o. Moreover, it is easy to check that $\Gamma(\mathcal{S}) = \Gamma(\mathcal{S}^o)$. In fact, the portion of the stalk space outside S^o is completely extraneous to the structure of global sections, so that the hypothesis that $\Gamma(\mathcal{S})$ be a subdirect product of the stalks is no real restriction at all (see Theorem 2.29, Corollary 2.33 and Theorems 2.34, 2.36).

For a second example we reformulate Theorem 2.20:

$$\mathcal{B} \subseteq_{\mathcal{T}} C_{\mathcal{T}}(I, \mathfrak{A}) \text{ iff } \mathcal{B} = C_{\mathcal{T}}(I, A) \cap \Pi\langle \pi_i(B) \mid i \in I\rangle.$$

Finally we shall discuss restrictions of sheaves. Let $\mathcal{S} = \langle S, \eta, I\rangle$ be a sheaf of structures and let $J \subseteq I$. We define the <u>restriction of</u> $\mathcal{S}$ <u>to</u>

J, denoted by $\mathcal{S} \mid J$, to be the sheaf $\langle S', \eta', J\rangle$, where

$$S' = \{a \in S \mid \eta(a) \in J\}$$

and η' is the restriction of η to S'.

Correspondingly, let $\mathcal{B} \subseteq \Pi\langle \mathfrak{A}_i \mid i \in I\rangle$ and let $J \subseteq I$.

Lemma 2.35 For every formula φ and every $x \in (\Pi A_i)^n$,

$$[\![\varphi(\pi_J(x))]\!] = [\![\varphi(x)]\!] \cap J. \blacksquare$$

Now suppose $\mathcal{B}$ is $\mathfrak{T}$-global and let $\mathfrak{T} \mid J$ be the relative topology on J. Then by Lemma 2.35, $\mathfrak{T} \mid J$ contains the equalizer topology induced by $\pi_J(\mathcal{B})$ and we define

$$\mathcal{B} \mid_{\mathfrak{T}} J = \Gamma(\pi_J(\mathcal{B}), \mathfrak{T} \mid J).$$

Theorem 2.36 Suppose $\mathcal{S} = \langle S, \eta, I\rangle$ is a sheaf of structures, where $\Gamma(\mathcal{S})$ is a subdirect product of the stalks. If $\mathfrak{T}$ is the topology on I and $J \subseteq I$ then

$$\Gamma(\mathcal{S} \mid J) = \Gamma(\mathcal{S}) \mid_{\mathfrak{T}} J.$$

Proof: Since $\Gamma(\mathcal{S})$ is a subdirect product of $\{\mathcal{X}_i \mid i \in I\}$, $\Gamma(\mathcal{S} \mid J)$ is a subdirect product of $\{\mathcal{X}_j \mid j \in J\}$. Now let $x \in \Gamma(\mathcal{S} \mid J)$. Then $x \in E\langle\pi_j(\Gamma(\mathcal{S})) \mid j \in J\rangle$. Since $\pi_J(\Gamma(\mathcal{S})) \subseteq \Gamma(\mathcal{S} \mid J)$, by Lemma 2.23, for every $y \in \Gamma(\mathcal{S})$, $E(x, \pi_J(y)) \in \mathfrak{T} \mid J$. Thus $x \in \Gamma(\pi_J(\Gamma(\mathcal{S})), \mathfrak{T} \mid J)$.

Next, since $\Gamma(\mathcal{S})$ is a subdirect product of $\{\mathcal{X}_i \mid i \in I\}$,

$$\{y(U) \mid y \in \Gamma(\mathcal{S}),\ U \in \mathfrak{T}\}$$

is a basis for the topology on S (consult the proof of Lemma 2.26). Moreover, for each $y \in \Gamma(\mathcal{S})$ and $U \in \mathfrak{T}$,

$$y(U) \cap \{a \in S \mid \eta(a) \in J\} = y(U \cap J).$$

Thus

$$\{\pi_J(y)(U \cap J) \mid y \in \Gamma(\mathcal{S}),\ U \in \mathfrak{T}\}$$

is a basis for the topology on $\{a \in S \mid \eta(a) \in J\}$. Now let $x \in \Gamma(\pi_J(\Gamma(\mathcal{S})), \mathfrak{T} \mid J)$. Then for every $y \in \Gamma(\mathcal{S})$, $E(x, \pi_J(y)) \in \mathfrak{T} \mid J$. Thus, by Lemma 2.25, for every $y \in \Gamma(\mathcal{S})$ and every $U \in \mathfrak{T}$,

$$x^{-1}(\pi_J(y)(U \cap J)) = U \cap E(x, \pi_J(y)) \cap J \in \mathfrak{T} \mid J$$

and therefore $x \in \Gamma(\mathcal{S} \mid J)$. $\blacksquare$

Theorem 2.36 establishes that $\mid_{\mathfrak{T}}$ is the correct analogue of sheaf restriction.

We shall conclude this section by showing how to "remove redundancies" from a global subdirect product. Suppose $\mathcal{B} \subseteq \Pi\langle \mathfrak{A}_i \mid i \in I\rangle$ is $\mathfrak{T}$-global and $J \subseteq I$ is a complete $\mathcal{B}$-transversal. For each $i \in I$, let $\rho(i)$ be the unique $j \in J$ such that $i \sim j(\mathcal{B})$. Then for each $i \in I$ there exists an isomorphism f_i from $\pi_{\rho(i)}(\mathcal{B})$ onto $\pi_i(\mathcal{B})$ such that for all $x \in B$,

$$x(i) = f_i(x(\rho(i))).$$

If we endow J with the induced quotient topology $\mathfrak{T}/\rho$ then $\rho : I \to J$ is a <u>continuous</u> <u>retraction</u>, that is $\rho(j) = j$ for every $j \in J$.

<u>Lemma</u> 2.37 If $\mathcal{B} \subseteq \Pi\langle \mathfrak{A}_i \mid i \in I\rangle$ is an algebra and $J \subseteq I$ is a complete $\mathcal{B}$-transversal then for every $x, y \in B$,

$$\rho^{-1}(E(\pi_J(x), \pi_J(y))) = E(x, y).$$

<u>Proof</u>: By Lemma 1.13, $\rho(i) \in E(x, y)$ if and only if $i \in E(x, y)$. ∎

<u>Lemma</u> 2.38 Suppose $\mathcal{B} \subseteq \Pi\langle \mathfrak{A}_i \mid i \in I\rangle$ is an algebra. Then $\mathcal{B}$ is irredundant if and only if the equalizer topology induced by $\mathcal{B}$ is T_o.

<u>Proof</u>: Suppose $\mathcal{B}$ is irredundant and $i \neq j$. By Lemma 1.13 there exist $x, y \in B$ such that exactly one of i and j belongs to $E(x, y)$. Thus the equalizer topology induced by $\mathcal{B}$ is T_o. Conversely, suppose the equalizer topology induced by $\mathcal{B}$ is T_o and $i \sim j(\mathcal{B})$. By Lemma 1.13, for every $x, y \in B$, $i \in E(x, y)$ if and only if $j \in E(x, y)$. It follows that $i = j$. ∎

<u>Theorem</u> 2.39 Suppose an algebra $\mathcal{B} \subseteq \Pi\langle \mathfrak{A}_i \mid i \in I\rangle$ is $\mathfrak{T}$-global and $J \subseteq I$ is a complete $\mathcal{B}$-transversal. Then $\mathcal{B} \cong \pi_J(\mathcal{B})$, where $\pi_J(\mathcal{B}) \subseteq \Pi\langle \mathfrak{A}_j \mid j \in J\rangle$ is $\mathfrak{T}/\rho$-global and J is a T_o-space. Moreover, if $E(x, y)$ is clopen for every $x, y \in B$ then J is Hausdorff.

<u>Proof</u>: By Lemma 2.37, $\mathfrak{T}/\rho$ contains the equalizer topology induced by $\pi_J(\mathcal{B})$. Suppose $x \in \Pi\langle \pi_j(B) \mid j \in J\rangle$ and for every $y \in B$, $E(x, \pi_J(y)) \in \mathfrak{T}/\rho$. Define for $i \in I$

$$z(i) = f_i(x(\rho(i)))$$

Then for every $y \in B$,

$$\begin{aligned} \rho(i) \in E(x, \pi_J(y)) \quad &\text{iff} \quad x(\rho(i)) = y(\rho(i)) \\ &\text{iff} \quad f_i(x(\rho(i))) = f_i(y(\rho(i))) \end{aligned}$$

$$\text{iff} \quad z(i) = y(i)$$

Thus

$$\rho^{-1}(E(x, \pi_J(y))) = E(z, y) \in \mathfrak{T}$$

because ρ is continuous. Thus $z \in B$ and therefore $x = \pi_J(z) \in \pi_J(B)$. This establishes that $\pi_J(\mathcal{B})$ is $\mathfrak{T}/\rho$-global. By Lemma 1.14 and 2.38, J is a T_o-space. Finally, if $E(x, y)$ is clopen for every $x, y \in B$ then, by Lemma 2.37, J is Hausdorff. ∎

Theorem 2.39 tells us that the index space of a $\mathfrak{T}$-global subdirect product may always be assumed to be T_o. Moreover, if equalizers are clopen, then the index space may be assumed to be Hausdorff.

3. THE HULL-KERNEL TOPOLOGY

To see whether a subdirect product $\mathcal{B} \subseteq \Pi\langle \mathfrak{A}_i \mid i \in I\rangle$ is global we have to consider the equalizer topology induced by $\mathcal{B}$. As indicated before, this topology is hopelessly elusive unless it is explicitly given in some other way which tells us more directly what the open sets are. For example, how could we hope to verify condition (ii) of Theorem 2.28: If $x \in \Pi A_i$ and if for every $y \in B$, $E(x, y)$ <u>is a union of finite intersections of equalizers from</u> B, then $x \in B$? Now in Section 2 we have seen three cases where a "natural" topology $\mathcal{T}$ was <u>given in advance</u>, two of them led to trivial global subdirect products (<u>diagonal</u> subdirect products and <u>direct</u> products) and one led to <u>weak direct</u> products, and in particular to <u>direct sums</u>. In this section we shall discuss a "natural" topology which may arise from a given subdirect representation. The hull-kernel topology was first introduced for rings by Jacobson [21] and has been widely used for sheaf representations of rings. We shall review some of the details in Section 7. It turns out, however, that the hull-kernel topology arises in a much more general setting, and that many global subdirect representations involve the hull-kernel topology in one form or another. Although the definition of the hull-kernel topology and some of its properties can be generalized to structures with relations, we presently see no useful applications of such a generalization. Accordingly, <u>throughout this section all structures are presumed to be algebras</u>.

Suppose $\mathcal{B} \subseteq \Pi\langle \mathfrak{A}_i \mid i \in I\rangle$ is <u>subdirect</u>. For each $F \subseteq I$ define

$$\overline{F} = \{i \in I \mid \theta_F^{\mathcal{B}} \subseteq \theta_i^{\mathcal{B}}\}.$$

<u>Lemma</u> 3.1 (i) $\overline{\emptyset} = \emptyset$ if and only if every $\mathfrak{A}_i$ is non-trivial.

(ii) $\overline{F} = \bigcap\{E(x, y) \mid x, y \in B \text{ and } F \subseteq E(x, y)\}$.

From the characterization of the operation $F \to \overline{F}$ given in Lemma 3.1 we obtain four consequences without referring to the original definition of this operation or the definition of equalizers:

<u>Lemma</u> 3.2 (i) $F \subseteq \overline{F}$

(ii) $\overline{\overline{F}} = \overline{F}$

(iii) If $F \subseteq G$ then $\overline{F} \subseteq \overline{G}$. ∎

Now let

$$\mathcal{R} = \{F \subseteq I \mid F = \overline{F}\}.$$

<u>Lemma</u> 3.3 $\langle \mathcal{R}, \wedge, \vee \rangle$ is a complete lattice, where for each $\mathfrak{X} \subseteq \mathcal{R}$,

$$\bigvee \mathfrak{X} = \overline{\bigcup \mathfrak{X}}$$
$$\bigwedge \mathfrak{X} = \bigcap \mathfrak{X}. \blacksquare$$

<u>Lemma</u> 3.4 The following are equivalent:

(i) $F \to \overline{F}$ is a topological closure operation.

(ii) $\emptyset = \overline{\emptyset}$ and for all $F, G \subseteq I$, $\overline{F \cup G} = \overline{F} \cup \overline{G}$.

(iii) $\emptyset = \overline{\emptyset}$ and for all $F, G \in \mathcal{R}$, $F \vee G = F \cup G$, that is, $\mathcal{R}$ is a ring of subsets of I.

<u>Proof</u>: Use Lemma 3.2. $\blacksquare$

We say $\mathcal{B}$ <u>induces a hull-kernel topology on</u> I if $F \to \overline{F}$ is a topological closure operation, and in this case the topology is called the <u>hull-kernel topology</u> on I (induced by $\mathcal{B}$).

<u>Corollary</u> 3.5 If $\mathcal{B}$ induces a hull-kernel topology on I then the hull-kernel topology has basis $\{D(x, y) \mid x, y \in B\}$. Conversely, if $\{D(x, y) \mid x, y \in B\}$ is a basis for some topology on I then $\mathcal{B}$ induces a hull-kernel topology on I.

<u>Proof</u>: The first part of the assertion follows from Lemma 3.1. To prove the second part, suppose $\{E(x, y) \mid x, y \in B\}$ is a basis of <u>closed</u> sets for a topology on I. Then $\emptyset = \overline{\emptyset}$ and for every $x, y, z, w \in B$, if $i \notin E(x, y) \cup E(z, w)$ then there exist $u, v \in B$ such that $E(x, y) \cup E(z, w) \subseteq E(u, v)$ and $i \notin E(u, v)$. It follows that $E(x, y) \cup E(z, w) \in \mathcal{R}$ whenever $x, y, z, w \in B$. Now let $F, G \in \mathcal{R}$, where by Lemma 3.1

$$F = \bigcap_{\xi \in \Delta} E(x_\xi, y_\xi) \quad \text{and} \quad G = \bigcap_{\eta \in \Sigma} E(z_\eta, w_\eta)$$

Then

$$F \cup G = \bigcap \{E(x_\xi, y_\xi) \cup E(z_\eta, w_\eta) \mid \xi \in \Delta, \eta \in \Sigma\}$$

and therefore, by Lemma 3.3, $F \cup G \in \mathcal{R}$. This establishes (iii) of Lemma 3.4. $\blacksquare$

The crucial issue for applications of course is to obtain non-trivial subdirect representations which induce a hull-kernel topology. To determine which algebras $\mathfrak{A}$ have a non-trivial subdirect representation inducing a hull-kernel topology we need a few definitions. Let Δ be a set of congruences on $\mathfrak{A}$ and for each $\Sigma \subseteq \Delta$ define

$$\overline{\Sigma} = \{\delta \in \Delta \mid \quad \Sigma \subseteq \delta\}.$$

Now define the canocical mapping $g:A \to \Pi\langle A/\delta \mid \delta \in \Delta\rangle$ by

$$g(x)(\delta) = x/\delta.$$

It is easy to check that $\Sigma \to \overline{\Sigma}$ is the closure operation on Δ induced by $g(\mathfrak{A})$. We say that $\mathfrak{A}$ <u>induces a hull-kernel topology on</u> Δ if $\Sigma \to \overline{\Sigma}$ is a topological closure operation and in this case the topology is called the hull-kernel topology on Δ.

<u>Corollary</u> 3.6 If Δ is a set of congruences on $\mathfrak{A}$ then $\mathfrak{A}$ induces a hull-kernel topology on Δ if and only if Δ does not contain the universal congruence and for every $\Sigma, \Gamma \subseteq \Delta$, $\overline{\Sigma \cup \Gamma} = \overline{\Sigma} \cup \overline{\Gamma}$.

<u>Proof</u>: Use Lemma 3.4. ∎

A congruence $\delta \in \Delta$ is called Δ-<u>irreducible</u> if δ is not the universal congruence and for every $\Sigma, \Gamma \subseteq \Delta$, if $(\bigcap \Sigma) \cap (\bigcap \Gamma) \subseteq \delta$ then $\bigcap \Sigma \subseteq \delta$ or $\bigcap \Gamma \subseteq \delta$. δ is called <u>irreducible</u> if δ is irreducible with respect to the set of all congruences on $\mathfrak{A}$.

<u>Theorem</u> 3.7 If Δ is a set of congruences on $\mathfrak{A}$ then $\mathfrak{A}$ induces a hull-kernel topology on Δ if and only if every member of Δ is Δ-irreducible.

<u>Proof</u>: Suppose $\mathfrak{A}$ induces a hull-kernel topology on Δ and let $\Sigma, \Gamma \subseteq \Delta$, where $(\bigcap \Sigma) \cap (\bigcap \Gamma) \subseteq \delta \in \Delta$. Then $\bigcap (\Sigma \cup F) \subseteq \delta$ and therefore $\delta \in \overline{\Sigma \cup \Gamma}$. By Corollary 3.6, $\delta \in \overline{\Sigma} \cup \overline{\Gamma}$ and therefore $\bigcap \Sigma \subseteq \delta$ or $\bigcap \Gamma \subseteq \delta$. This proves one direction of the assertion and the other is proved by reversing the argument. ∎

<u>Corollary</u> 3.8 $\mathfrak{A}$ induces a hull-kernel topology on any set of irreducible congruences. ∎

Now we can characterize algebras which have a non-trivial subdirect representation inducing a hull-kernel topology.

<u>Theorem</u> 3.9 Suppose Δ is a set of congruences on $\mathfrak{A}$. Then the canonical mapping $g:A \to \Pi\langle A/\delta \mid \delta \in \Delta\rangle$ is a non-trivial subdirect representation of

$\mathfrak{A}$ inducing a hull-kernel topology on Δ if and only if every member of Δ is Δ-irreducible, $\Phi^{\mathfrak{A}} \notin \Delta$ and $\bigcap \Delta = \Phi^{\mathfrak{A}}$.

Proof: Use Theorem 3.7. ∎

Corollary 3.10 Suppose Δ is a set of irreducible congruences on $\mathfrak{A}$. Then the canonical mapping $g: A \to \Pi\langle A/\delta \mid \delta \in \Delta\rangle$ is a subdirect representation of $\mathfrak{A}$ inducing a hull-kernel topology on Δ if and only if $\bigcap \Delta = \Phi^{\mathfrak{A}}$. ∎

$\mathfrak{A}$ is called semi irreducible if $\mathfrak{A}$ is non-trivial and the intersection of the set of all irreducible congruences of $\mathfrak{A}$ is the identity congruence.

Corollary 3.11 If $\mathfrak{A}$ is semi irreducible and Δ is the set of all irreducible congruences on $\mathfrak{A}$ then the canonical mapping $g: A \to \Pi\langle A/\delta \mid \delta \in \Delta\rangle$ is a subdirect representation of $\mathfrak{A}$ inducing a hull-kernel topology on Δ. ∎

The significance of the hull-kernel topology for sheaf representations of rings is due to the fact that a "natural" class of rings (namely the semi-prime rings) consists of semi irreducible rings. However, this appears to be a specifically ring theoretic phenomenon. To obtain semi irreducible algebras in a more general setting (which, interestingly enough, does not apply to semi-prime rings) we have to relate our notion of irreducibility to some well-known notions from universal algebras. A congruence θ on $\mathfrak{A}$ is called meet irreducible if θ is not the universal congruence and for all congruences ψ and χ on $\mathfrak{A}$, if $\theta = \psi \cap \chi$ then $\theta = \psi$ or $\theta = \chi$. θ is called completely meet irreducible if θ is not the universal congruence and for every family $\{\chi_i \mid i \in I\}$ of congruences on $\mathfrak{A}$, if $\theta = \bigcap_{i \in I} \chi_i$ then $\theta = \chi_i$ for some $i \in I$. Notice that $\Phi^{\mathfrak{A}}$ is irreducible if and only if $\Phi^{\mathfrak{A}}$ is meet irreducible. An algebra $\mathfrak{A}$ is called subdirectly indecomposable if $\mathfrak{A}$ is non-trivial and $\Phi^{\mathfrak{A}}$ is (meet) irreducible. In particular, a subdirectly indecomposable algebra is semi irreducible.

Lemma 3.12 (i) θ is meet irreducible if and only if $\mathfrak{A}/\theta$ is subdirectly indecomposable.
(ii) θ is completely meet irreducible if and only if $\mathfrak{A}/\theta$ is subdirectly irreducible. ∎

Lemma 3.13 (i) If θ is irreducible then θ is meet irreducible.
(ii) If $\mathfrak{A}$ is congruence distributive then θ is irreducible

and only if θ is meet irreducible.

Proof: (i) is obvious. To prove (ii), assume $\mathfrak{A}$ is congruence distributive and θ is meet irreducible. Suppose $\psi \cap \chi \subseteq \theta$. Then

$$\theta = (\psi \cap \chi) \vee \theta = (\psi \vee \theta) \cap (\chi \vee \theta)$$

and therefore $\theta = \psi \vee \theta$ or $\theta = \chi \vee \theta$. Thus $\psi \subseteq \theta$ or $\chi \subseteq \theta$. ∎

Corollary 3.14 A congruence distributive algebra is semi irreducible.

Proof: By Lemma 3.13, every completely meet irreducible congruence is irreducible. By Birkhoff's Theorem, the intersection of the set of all completely meet irreducible congruences is the identity congruence. ∎

A congruence distributive algebra has a subdirect representation by subdirectly irreducibles inducing a hull-kernel topology. In a sense the converse is true. To see this, we return to the setting where $\mathfrak{B} \subseteq \Pi\langle \mathfrak{A}_i \mid i \in I\rangle$ is subdirect. The crucial observation is that there is a close relationship between the ring $\mathcal{R}$ of closed subsets of I and the corresponding projection congruences of $\mathfrak{B}$.

Lemma 3.15 For any $F, G \subseteq I$,

(i) $\theta_F^{\mathfrak{B}} \subseteq \theta_G^{\mathfrak{B}}$ if and only if $\overline{G} \subseteq \overline{F}$.

(ii) $\theta_F^{\mathfrak{B}} = \theta_G^{\mathfrak{B}}$ if and only if $\overline{F} = \overline{G}$.

(iii) $\theta_F^{\mathfrak{B}} \cap \theta_G^{\mathfrak{B}} = \theta_{F\cup G}^{\mathfrak{B}}$.

(iv) $\theta_\emptyset^{\mathfrak{B}} = \mathbb{1}^{\mathfrak{B}}$ and $\theta_I^{\mathfrak{B}} = \mathbb{0}^{\mathfrak{B}}$. ∎

To obtain information about the lattice $\langle\mathcal{R}, \wedge, \vee\rangle$ of hull-kernel closed sets form the congruence lattice of $\mathfrak{B}$ via the injection $F \to \theta_F^{\mathfrak{B}}$ we have to look for the missing property of a lattice-antiembedding.

Lemma 3.16 The projections of $\mathfrak{B}$ determine a sublattice of the congruence lattice of $\mathfrak{B}$ if and only if for every $F, G \subseteq I$,

$$\theta_F^{\mathfrak{B}} \vee \theta_G^{\mathfrak{B}} = \theta_{\overline{F}\cap\overline{G}}^{\mathfrak{B}} .$$

Proof: Suppose the projections of $\mathfrak{B}$ determine a sublattice of the congruence lattice of $\mathfrak{B}$ and let $F, G \subseteq I$. Then by Lemma 3.15,

$$\theta_F^{\mathcal{L}} \vee \theta_G^{\mathcal{L}} = \theta_{\overline{F}}^{\mathcal{L}} \vee \theta_{\overline{G}}^{\mathcal{L}} \subseteq \theta_{\overline{F \cap G}}^{\mathcal{L}}$$

Now suppose $\theta_F^{\mathcal{L}} \vee \theta_G^{\mathcal{L}} = \theta_H^{\mathcal{L}}$. By Lemma 3.15,

$$\overline{H} \subseteq \overline{F} \cap \overline{G} = \overline{\overline{F} \cap \overline{G}}$$

and therefore $\theta_{\overline{F \cap G}}^{\mathcal{L}} \subseteq \theta_H^{\mathcal{L}}$. This establishes one direction of the assertion and the other follows from Lemma 3.15. ∎

Corollary 3.17 The mapping $F \in \mathcal{R} \to \theta_F^{\mathcal{L}}$ is an antiembedding of $\mathcal{R}$ into the congruence lattice of $\mathcal{L}$ if and only if the projections of $\mathcal{L}$ determine a sublattice of the congruence lattice of $\mathcal{L}$.

Proof: Use Lemma 3.15 and 3.16. ∎

Corollary 3.18 If $\mathcal{L}$ induces a hull-kernel topology on I and the projections of $\mathcal{L}$ determine a sublattice of the congruence lattice of $\mathcal{L}$ then this sublattice is distributive.

Proof: By Corollary 3.4, $\mathcal{R}$ is a distributive lattice. Now use Corollary 3.17. ∎

In general one only can tell that the projections determine a sublattice of the congruence lattice in case every congruence is a projection. $\mathcal{L}$ is called fully expanded if every congruence on $\mathcal{L}$ is a projection.

Corollary 3.19 If $\mathcal{L}$ induces a hull-kernel topology on I and is fully expanded then $\mathcal{L}$ is congruence distributive and every factor $\mathfrak{A}_i$ is subdirectly indecomposable.

Proof: By Corollary 3.18, $\mathcal{L}$ is congruence distributive and by Theorem 3.9, every $\theta_i^{\mathcal{L}}$ is irreducible. Thus by Lemma 3.13, every factor $\mathfrak{A}_i$ is subdirectly indecomposable. ∎

Lemma 3.20 Let Δ be a set of congruences on $\mathfrak{A}$. Then the canonical mapping $g:A \to \Pi\langle A/\delta \mid \delta \in \Delta\rangle$ is a fully expanded subdirect representation of $\mathfrak{A}$ if and only if Δ contains all completely meet irreducible congruences on $\mathfrak{A}$.

Proof: Assume $g:A \to \Pi\langle A/\delta \mid \delta \in \Delta\rangle$ is a fully expanded subdirect representation of $\mathfrak{A}$ and let θ be a completely meet irreducible congruence on $\mathfrak{A}$. Let $\mathfrak{B} = g(\mathfrak{A})$. Then there exists $\Sigma \subseteq \Delta$ such that $\theta^{\mathfrak{B}}_{\Sigma}$ is completely meet irreducible and for every $x, y \in A$, $x\theta y$ if and only if $g(x)\theta^{\mathfrak{B}}_{\Sigma} g(y)$. Since

$$\theta^{\mathfrak{B}}_{\Sigma} = \bigcap_{\delta\in\Sigma} \theta^{\mathfrak{B}}_{\delta}$$

there exists $\delta \in \Sigma$ such that $\theta^{\mathfrak{B}}_{\Sigma} = \theta^{\mathfrak{B}}_{\delta}$. Since for every $x, y \in A$, $x\delta y$ if and only if $g(x)\theta^{\mathfrak{B}}_{\delta} g(y)$, $\theta = \delta$. Thus $\theta \in \Delta$.

Conversely, assume Δ contains all completely meet irreducible congruences on $\mathfrak{A}$. By Birkhoff's Theorem, $g:A \to \Pi\langle A/\delta \mid \delta \in \Delta\rangle$ is a subdirect representation of $\mathfrak{A}$. Let $\mathfrak{B} = g(\mathfrak{A})$ and let θ be a congruence on $\mathfrak{B}$. If $\theta = \mathbb{1}^{\mathfrak{B}}$ then $\theta = \theta^{\mathfrak{B}}_{\emptyset}$. Otherwise, let $\mathfrak{C} = \mathfrak{B}/\theta$ and let $\{\psi_i \mid i \in I\}$ be the set of all completely meet irreducible congruences on $\mathfrak{C}$. It is easy to see that $\{\theta^{\mathfrak{B}}_{\delta} \mid \delta \in \Delta\}$ contains all completely meet irreducible congruences on $\mathfrak{B}$. Thus for each $i \in I$ there exists $\delta_i \in \Delta$ such that for all $x, y \in B$,

$$x\theta^{\mathfrak{B}}_{\delta_i} y \quad \text{iff} \quad (x/\theta)\psi_i(y/\theta).$$

We claim that

$$\theta = \bigcap_{i\in I} \theta_{\delta_i}$$

Indeed, suppose $x\theta^{\mathfrak{B}}_{\delta_i} y$ for all $i \in I$. Then $(x/\theta)\psi_i(y/\theta)$ for all $i \in I$ and therefore, by Birkhoff's Theorem, $x\theta y$. This establishes that $\mathfrak{B}$ is fully expanded. ∎

Corollary 3.21 Every non-trivial algebra is isomorphic to a fully expanded subdirect product of subdirectly irreducible algebras. ∎

Theorem 3.22 If $\mathfrak{A}$ is a non-trivial algebra then the following are equivalent.

(i) $\mathfrak{A}$ is congruence distributive.

(ii) $\mathfrak{A}$ has a fully expanded subdirect representation by subdirectly irreducibles inducing a hull-kernel topology.

(iii) $\mathfrak{A}$ has a fully expanded subdirect representation inducing a hull-kernel topology.

Proof: Assume (i) and let Δ be the set of all completely meet irreducible congruences on $\mathfrak{A}$. By Lemma 3.20, the canonical mapping $g:A \to \Pi\langle A/\delta \mid \delta \in \Delta\rangle$ is a fully expanded subdirect representation of $\mathfrak{A}$. By Lemma 3.13, every member of Δ is irreducible and by Corollary 3.8, $\mathfrak{A}$ induces a hull-kernel topology on Δ. This establishes (ii). Finally, (ii) implies (iii) trivially and (iii) implies (i) by Corollary 3.19. ∎

It would be interesting to know whether Theorem 3.22 can be further strengthened:

<u>Problem</u> 3.23 Suppose $\mathcal{V}$ is a variety where every $\mathfrak{A} \in \mathcal{V}$ has a subdirect representation by subdirectly irreducibles inducing a hull-kernel topology. Is $\mathcal{V}$ congruence distributive?

We can, however, improve Lemma 3.13.

<u>Theorem</u> 3.24 If $\mathfrak{A}$ is a non-trivial algebra then the following are equivalent:

(i) $\mathfrak{A}$ is congruence distributive.

(ii) Every meet irreducible congruence on $\mathfrak{A}$ is irreducible.

(iii) Every completely meet irreducible congruence on $\mathfrak{A}$ is irreducible.

<u>Proof</u>: (i) implies (ii) by Lemma 3.13 and (ii) implies (iii) trivially. Finally assume (iii) and let Δ be the set of completely meet irreducible congruences on $\mathfrak{A}$. By Lemma 3.20, the canonical mapping $g:A \to \Pi\langle A/\delta \mid \delta \in \Delta\rangle$ is a fully expanded subdirect representation of $\mathfrak{A}$ and by Corollary 3.8, $\mathfrak{A}$ induces a hull-kernel topology on Δ. (i) follows from Theorem 3.22. ∎

We again return to the setting where $\mathfrak{B} \subseteq \Pi\langle \mathfrak{A}_i \mid i \in I\rangle$ is subdirect. If $\mathfrak{B}$ induces a hull-kernel topology on I and is fully expanded then we obtain a great deal of information about the hull-kernel topology from the congruence lattice of $\mathfrak{B}$. In this case, by Corollary 3.17, the mapping $F \in \mathcal{R} \to \theta_F^{\mathfrak{B}}$ is an antiisomorphism between the ring of hull-kernel closed sets and the congruence lattice of $\mathfrak{B}$. Moreover, since every congruence on $\mathfrak{B}$ is a projection, $\theta^{\mathfrak{B}}_{E(x,\ y)} = \theta^{\mathfrak{B}}(x,\ y)$ for every $x,\ y \in B$.

<u>Lemma</u> 3.25 Suppose $\mathfrak{B}$ induces a hull-kernel topology on I and is fully expanded. Then an open set G is compact if and only if there exist $r < \omega$ and $x_p,\ y_p \in B$ for each $p < r$ such that $G = \bigcup_{p<r} D(x_p,\ y_p)$.

<u>Proof</u>: The mapping $G \to \theta^{\mathfrak{B}}_{I\ -\ G}$ is an isomorphism between the ring of hull-kernel open sets and the congruence lattice of $\mathfrak{B}$. Moreover, this mapping sends $D(x,\ y)$ to $\theta^{\mathfrak{B}}(x,\ y)$. The compact congruences on $\mathfrak{B}$ are exactly the finite joins of principal congruences. The assertion follows. ∎

<u>Corollary</u> 3.26 If $\mathfrak{B}$ induces a hull-kernel topology on I and is fully expanded then the hull-kernel topology is locally compact. Moreover,

the following are equivalent:

(i) The hull-kernel topology is compact.

(ii) There exist $r < \omega$ and $x_p, y_p \in B$ for each $p < r$ such that $I = \bigcup_{p<r} D(x_p, y_p)$.

(iii) The universal congruence on $\mathcal{L}$ is compact. ∎

<u>Corollary</u> 3.27 If $\mathcal{L}$ induces a hull-kernel topology on I and is fully expanded then for any $x, y \in B$, $\mathbb{1}^{\mathcal{L}} = \theta^{\mathcal{L}}(x, y)$ if and only if $I = D(x, y)$. ∎

As indicated before, the hull-kernel topology does not always come from congruence distributively. For example, a semi-prime ring has a subdirect representation by prime rings inducing a hull-kernel topology. However, a semi-prime ring is not necessarily congruence distributive, a prime ring is not necessarily subdirectly irreducible, and the projections of the subdirect representation do not necessarily determine a sublattice of the congruence lattice. We have not been able to discover a universal algebraic pattern in this situation, but we can give a few technical results which will be helpful later. The first one gives us compactness of the hull-kernel topology.

<u>Lemma</u> 3.28 Suppose $\mathcal{L}$ induces a hull-kernel topology on I. If the universal congruence on $\mathcal{L}$ is compact and $\{\theta_i^{\mathcal{L}} \mid i \in I\}$ contains all maximal congruences on $\mathcal{L}$ then the hull-kernel topology is compact.

<u>Proof</u>: If $\mathcal{S} \subseteq \mathcal{R}$ then for every $i \in I$,

$$i \in \bigcap \mathcal{S} \quad \text{iff} \quad \bigvee_{F \in \mathcal{S}} \theta_F^{\mathcal{L}} \subseteq \theta_i^{\mathcal{L}}.$$

Now suppose $\bigcap \mathcal{S} = \emptyset$. Then for every $i \in I$, $\bigvee_{F \in \mathcal{S}} \theta_F^{\mathcal{L}} \nsubseteq \theta_i^{\mathcal{L}}$. Since the universal congruence is compact, every proper congruence is contained in a maximal congruence. Thus by hypothesis, $\bigvee_{F \in \mathcal{S}} \theta_F^{\mathcal{L}} = \mathbb{1}^{\mathcal{L}}$ and therefore there exists finite $\mathcal{P} \subseteq \mathcal{S}$ such that $\bigvee_{F \in \mathcal{P}} \theta_F^{\mathcal{L}} = \mathbb{1}^{\mathcal{L}}$. Since each $\mathfrak{A}_i$ is non-trivial, $\bigvee_{F \in \mathcal{P}} \theta_F^{\mathcal{L}} \nsubseteq \theta_i^{\mathcal{L}}$ for all $i \in I$ and therefore $\bigcap \mathcal{P} = \emptyset$. ∎

The next result will play an important role in "patching" arguments over hull-kernel open sets.

<u>Lemma</u> 3.29 Suppose $\mathcal{L}$ induces a hull-kernel topology on I, where every proper congruence is contained in some $\theta_i^{\mathcal{L}}$. If $\{G_p \mid p < r\}$ is a

finite open covering of I then $\mathbb{1}^{\mathcal{L}} = \bigvee_{p<r} \theta^{\mathcal{L}}_{I-G}$.

Proof: For each congruence χ on $\mathcal{L}$ define

$$S(\chi) = \{i \in I \mid \chi \nsubseteq \theta^{\mathcal{L}}_i\}$$

and notice that

(i) for any two congruences ψ and χ on $\mathcal{L}$,

$$S(\psi) \cup S(\chi) = S(\psi \vee \chi);$$

(ii) for any open set G, $S(\theta^{\mathcal{L}}_{I-G}) = G$.

Thus

$$I = \bigcup_{p<r} G_p = \bigcup_{p<r} S(\theta^{\mathcal{L}}_{I-G_p}) = S(\bigvee_{p<r} \theta^{\mathcal{L}}_{I-G_p})$$

and it follows from our hypothesis that $\mathbb{1}^{\mathcal{L}} = \bigvee_{p<r} \theta^{\mathcal{L}}_{I-G_p}$. ∎

In applications the hull-kernel topology usually is T_o, whereas stronger separation properties are rather special. This is not accidental.

Lemma 3.30 Suppose $\mathcal{L}$ induces a hull-kernel topology on I. Then $\mathcal{L}$ is irredundant if and only if the hull-kernel topology is T_o.

Proof: Suppose $\mathcal{L}$ is irredundant and $i \neq j$. By Lemma 1.13 there exist $x, y \in B$ such that exactly one of i and j belong to $D(x, y)$. By Corollary 3.5, the hull-kernel topology is T_o. Conversely, suppose the hull-kernel topology is T_o and $i \sim j(\mathcal{L})$. By Lemma 1.13, for every $x, y \in B$, $i \in E(x, y)$ if and only if $j \in E(x, y)$. By Corollary 3.5, $i = j$. ∎

By Lemma 1.14 we can always "cut down" to a complete $\mathcal{L}$-transversal. This process neither disturbs the hull-kernel topology nor full expansion. At this point we should like to point out a subtle detail which is easily overlooked. If $\mathcal{L} \subseteq \Pi\langle \mathfrak{A}_i \mid i \in I\rangle$, $\mathcal{T}$ is a topology on I and $J \subseteq I$ is a complete $\mathcal{L}$-transversal then we can endow J either with the relative topology $\mathcal{T} \mid J$ or with the quotient topology $\mathcal{T}/\rho$, where $\rho: I \to J$ is the continuous retraction defined at the end of Section 2. Then we obtain $\mathcal{T}/\rho \subseteq \mathcal{T} \mid J$ because for every $F \subseteq J$, $F = \rho^{-1}(F) \cap J$. However, in general these two topologies are not the same!

Lemma 3.31 If $\mathcal{L} \subseteq \Pi\langle \mathfrak{A}_i \mid i \in I\rangle$ induces a hull-kernel topology on I and $J \subseteq I$ then $\pi_J(\mathcal{L}) \subseteq \Pi\langle \mathfrak{A}_j \mid j \in J\rangle$ induces a hull-kernel topology on J, which is the relative topology $\mathcal{T} \mid J$. If J is a complete

$\mathcal{L}$-transversal then the relative topology $\mathfrak{T} \mid J$ and the quotient topology $\mathfrak{T}/\rho$ are the same. ∎

4. PATCHING

In this section we shall establish the central results of this study. To begin with we return to the characterization of global subdirect products given in Corollary 2.3, and we make a simple observation: Many sheaf representations originate in a situation where a subdirect representation $\mathcal{B} \subseteq \Pi\langle \mathfrak{A}_i \mid i \in I\rangle$ and a topology $\mathfrak{T}$ on I are given and either one or both of the two conditions of Corollary 2.3 are actually violated. Now condition (ii) of this corollary has a strongly "infinitary" flavor and we should expect verification difficult unless circumstances are rather special. So we shall first restate Corollary 2.3 so that the "finitary" core of condition (ii) becomes visible.

Lemma 4.1 Suppose $\mathcal{B} \subseteq \Pi\langle \mathfrak{A}_i \mid i \in I\rangle$ and $\mathfrak{T}$ is a topology on I. Then $\mathcal{B}$ is $\mathfrak{T}$-global if and only if

(i) for every atomic formula φ and every $x \in B^n$, $[\![\varphi(x)]\!] \in \mathfrak{T}$;

(ii) (Unrestricted Patching) if $F_\zeta \in \mathfrak{T}$ and $x_\zeta \in B$ for each $\zeta \in \Delta$, where $I = \bigcup_{\zeta\in\Delta} F_\zeta$ and $F_\zeta \cap F_\eta \subseteq E(x_\zeta, x_\eta)$ whenever $\zeta, \eta \in \Delta$ then there exists $y \in B$ such that $F_\zeta \subseteq E(x_\zeta, y)$ for all $\zeta \in \Delta$.

Proof: Suppose $\mathcal{B}$ is $\mathfrak{T}$-global, and let $F_\zeta \in \mathfrak{T}$ and $x_\zeta \in B$ for each $\zeta \in \Delta$, where $I = \bigcup_{\zeta\in\Delta} F_\zeta$ and $F_\zeta \cap F_\eta \subseteq E(x_\zeta, x_\eta)$ whenever $\zeta, \eta \in \Delta$. Define $y \in \Pi\pi_i(B)$ by

$$y(i) = x_\zeta(i), \quad \text{where } i \in F_\zeta$$

and consider any $z \in B$. Then

$$E(y, z) = \bigcup_{\zeta\in\Delta} F_\zeta \cap E(y, x_\zeta) \cap E(x_\zeta, z)$$
$$= \bigcup_{\zeta\in\Delta} F_\zeta \cap E(x_\zeta, z) \in \mathfrak{T}$$

and $y \in B$. This establishes unrestricted patching.

Conversely, assume (i) and (ii) of the assertion and let $x \in \Pi\pi_i(B)$, where $E(x, y) \in \mathfrak{T}$ for all $y \in B$. Then for each $i \in I$ there exists $y_i \in B$ such that $i \in E(x, y_i)$. Thus $I = \bigcup_{i\in I} E(x, y_i)$ and clearly $E(x, y_i) \cap E(x, y_j) \subseteq E(y_i, y_j)$ whenever $i, j \in I$. Thus there exists $z \in B$ such that $E(x, y_i) \subseteq E(z, y_i)$ for all $i \in I$. It follows that $x = z$, and $\mathcal{B}$ is $\mathfrak{T}$-global. ∎

Lemma 4.1 was discovered independently by Weispfenning [38]. It is clear that the finitary core of unrestricted patching is finite patching: if $F_p \in \mathfrak{T}$ and $x_p \in B$ for each $p < r < \omega$, where $I = \bigcup_{p<r} F_p$ and $F_p \cap F_q \subseteq E(x_p, x_q)$ whenever $p, q < r$, then there exists $y \in B$ such that $F_p \subseteq E(y, x_p)$ for all $p < r$. Many sheaf representations arise from concrete subdirect representations where finite patching can be directly verified by some bare-hands effort. To obtain unrestricted patching usually some form of compactness enters the argument. The most peculiar features of sheaf representation occur when condition (i) of Lemma 4.1 is violated. Since in many cases the topology $\mathfrak{T}$ is the hull-kernel topology induced by $\mathfrak{B}$, and since this topology is predestined to violate condition (i) of Lemma 4.1 (see Corollary 3.5), this explains why many sheaf representations actually are rather peculiar. We shall now give a completely canonical construction in the setting of universal algebra which yields all sheaf representations starting from finite patching.

Let I be a non-empty set and let $\mathcal{R}$ be a family of subsets of I. $\mathcal{R}$ is called a dual ring of subsets of I if

(i) $I \in \mathcal{R}$;

(ii) if $F, G \in \mathcal{R}$ then $F \cap G$, $F \cup G \in \mathcal{R}$.

Notice, if $\mathcal{R}$ is a ring of subsets of I and

$$\breve{\mathcal{R}} = \{I - F \mid F \in \mathcal{R}\}$$

then $\breve{\mathcal{R}}$ is a dual ring of subsets of I.

Now let $\mathfrak{B} \subseteq \Pi\langle \mathfrak{A}_i \mid i \in I\rangle$ and let $\mathcal{R}$ be a dual ring of subsets of I. We say that $\mathfrak{B}$ patches over $\mathcal{R}$ if for every $r < \omega$, if $F_p \in \mathcal{R}$ and $x_p \in B$ for each $p < r$, where $I = \bigcup_{p<r} F_p$ and $F_p \cap F_q \subseteq E(x_p, x_q)$ whenever $p, q < r$, then there exists $y \in B$ such that $F_p \subseteq E(y, x_p)$ for all $p < r$. For technical reasons we shall frequently start with a ring $\mathcal{R}$ of subsets of I and consider patching over the dual ring $\breve{\mathcal{R}}$. In such a situation it is helpful to recall the following observation:

Lemma 4.2 Let $\mathfrak{B} \subseteq \Pi\langle \mathfrak{A}_i \mid i \in I\rangle$ and let $\mathcal{R}$ be a ring of subsets of I. Then $\mathfrak{B}$ patches over $\breve{\mathcal{R}}$ if and only if for every $r < \omega$, if $F_p \in \mathcal{R}$ and $x_p \in B$ for each $p < r$, where $\bigcap_{p<r} F_p = \emptyset$ and $D(x_p, x_q) \subseteq F_p \cup F_q$ whenever $p, q < r$, then there exists $y \in B$ such that $D(y, x_p) \subseteq F_p$ for all $p < r$. ∎

Notice that a structure may patch over a dual ring in quite a trivial fashion in case the dual ring is "small" or not "closely related" to the structure. This motivates our next definition. From now on let

$\mathcal{B} \subseteq \Pi\langle \mathfrak{A}_i \mid i \in I\rangle$ and let $\mathcal{R}$ be a ring of subsets of I. Intuitively we think of $\mathcal{R}$ a being a basis of closed sets for a topology $\mathfrak{T}(\mathcal{R})$ on I (See Section 1 on the Wallman compactification). This topology has the dual ring $\breve{\mathcal{R}}$ as a basis of open sets. $\mathcal{B}$ is called semi normal over $\mathcal{R}$ if

(i) $\mathcal{B}$ patches over $\breve{\mathcal{R}}$;

(ii) for every atomic formula φ and every $x \in B^n$, $[\![\neg\varphi(x)]\!]$ is $\mathfrak{T}(\mathcal{R})$-closed.

$\mathcal{B}$ is called normal over $\mathcal{R}$ if

(i) $\mathcal{B}$ patches over $\breve{\mathcal{R}}$;

(ii) for every atomic formula φ and every $x \in B^n$, $[\![\neg\varphi(x)]\!] \in \mathcal{R}$.

Lemma 4.3 If $\mathcal{B} \subseteq \Pi\langle \mathfrak{A}_i \mid i \in I\rangle$ is $\mathfrak{T}$-global then $\mathcal{B}$ is semi normal over every ring $\mathcal{R}$ of subsets of I which is a basis of closed sets for $\mathfrak{T}$. Moreover, $\mathcal{B}$ is normal over the ring of all closed sets.

Proof: Use Lemma 4.1. ∎

It is clear why compactness plays such a central role: it reduces unrestricted patching to (finite) patching.

Lemma 4.4 Suppose $\mathcal{B} \subseteq \Pi\langle \mathfrak{A}_i \mid i \in I\rangle$ and $\mathcal{R}$ is a ring of subsets of I. Let $\mathfrak{T}(\mathcal{R})$ be the topology on I with basis of closed sets $\mathcal{R}$. If $\mathcal{B}$ is semi normal over $\mathcal{R}$ and I is compact then $\mathcal{B}$ is $\mathfrak{T}(\mathcal{R})$-global.

Proof: Suppose $x \in \Gamma(B, \mathfrak{T}(\mathcal{R}))$. For each $i \in I$ there exists $y_i \in B$ such that $x(i) = y_i(i)$, that is $i \in E(x, y_i)$. Thus

$$I = \bigcup_{i \in I} E(x, y_i)$$

where $E(x, y_i)$ is open for all $i \in I$. Since $\breve{\mathcal{R}}$ is a basis for $\mathfrak{T}(\mathcal{R})$ and I is compact, there exist $r < \omega$ and $F_p \in \breve{\mathcal{R}}$, for each $p < r$, such that

$$I = \bigcup_{p<r} F_p$$

and for each $p < r$ there exists $i_p \in I$ such that $F_p \subseteq E(x, y_{i_p})$. Now clearly

$$F_p \cap F_q \subseteq E(x, y_{i_p}) \cap E(x, y_{i_q}) \subseteq E(y_{i_p}, y_{i_q}),$$

whenever $p, q < r$. Since $\mathcal{B}$ patches over $\breve{\mathcal{R}}$, there exists $z \in B$ such that $F_p \subseteq E(z, y_{i_p})$ for all $p < r$. It follows that $x = z$ and the assertion is proved. ∎

Patching is particulary simple over a field of subsets.

Lemma 4.5 Let $\mathcal{B} \subseteq \Pi\langle \mathfrak{A}_i \mid i \in I\rangle$ and let $\mathfrak{F}$ be a field of subsets of I. Then $\mathcal{B}$ patches over $\mathfrak{F}$ if and only if for every $x, y \in B$ and every $F \in \mathfrak{F}$, if

$$z(i) = \begin{cases} x(i) & \text{for } i \in F \\ y(i) & \text{for } i \notin F \end{cases}$$

then $z \in B$.

Proof: Assume the right-hand side of the assertion and let $r < \omega$, $F_p \in \mathfrak{F}$ and $x_p \in B$ for each $p < r$, where $I = \bigcup_{p<r} F_p$ and $F_p \cap F_q \subseteq E(x_p, x_q)$ whenever $p, q < r$. Since $\mathfrak{F}$ is a field there exist $G_p \in \mathfrak{F}$, for each $p < r$, such that $I = \bigcup_{p<r} G_p$, where $\{G_p \mid p < r\}$ is disjoint and $G_p \subseteq F_p$ for all $p < r$. First let

$$y(i) = x_p(i) \quad \text{if } i \in G_p$$

and then define

$$z_o = x_o$$

$$z_{p+1}(i) = \begin{cases} z_p(i) & \text{if } i \notin G_{p+1} \\ x_{p+1}(i) & \text{if } i \in G_{p+1} \end{cases}$$

Then $z_o(i) = y(i)$ for all $i \in G_o$. Next, suppose $z_p(i) = y(i)$ for all $i \in G_o \cup \ldots \cup G_p$. Let $i \in G_o \cup \ldots \cup G_{p+1}$ and distinguish two cases Case 1: $i \in G_o \cup \ldots \cup G_p$. Then $i \notin G_{p+1}$ and therefore $z_{p+1}(i) = z_p(i) = y(i)$. Case 2: $i \in G_{p+1}$. Then $z_{p+1}(i) = x_{p+1}(i) = y(i)$. Thus by induction, $z_{r-1} = y \in B$.

Now let $i \in F_p$. If $i \in G_q$ then $i \in F_p \cap F_q$ and therefore $y(i) = x_q(i) = x_p(i)$. This shows that $\mathcal{B}$ patches over $\mathfrak{F}$. ∎

A topological space is called zero-dimensional if it has a basis of clopen sets.

Corollary 4.6 (Burris and Werner) Suppose $\mathcal{B} \subseteq \Pi\langle \mathfrak{A}_i \mid i \in I\rangle$ and $\mathfrak{T}$ is a topology on I containing the equalizer topology induced by $\mathcal{B}$. If I is a compact zero-dimensional space then $\mathcal{B}$ is $\mathfrak{T}$-global if and only if for every $x, y \in B$ and every clopen F, if

$$z(i) = \begin{cases} x(i) & \text{for } i \in F \\ y(i) & \text{for } i \notin F \end{cases}$$

then $z \in B$.

Proof: Use Lemmas 4.3, 4.4 and 4.5. ∎

(Semi) normal subdirect products are natural generalizations of global subdirect products which capture the finitary aspects of "global tightness". Of course, in the absence of compactness (semi) normalcy does not in general imply globality. However, we are always able to globalize a semi normal subdirect product at the expense of obtaining new factors. More generally, if a subdirect product patches over a dual ring of subsets then we can always globalize it at the expense of "blowing up" the old factors and obtaining new factors. To work out the details, let $\mathcal{B} \subseteq \Pi\langle \mathfrak{A}_i \mid i \in I\rangle$ be subdirect and let $\mathcal{R}$ be a ring of subsets of I. Assume that $\mathcal{B}$ patches over $\breve{\mathcal{R}}$ and let $\mathfrak{T}(\mathcal{R})$ be the topology with basis of closed sets $\mathcal{R}$. Using the notation of Section 1, let J be the set of prime filters on $\mathcal{R}$. For $\mathcal{U} \in J$ and $x, y \in B$ define

$$x \equiv y(\mathcal{U}) \text{ iff for some } F \in \mathcal{R},\ F \notin \mathcal{U} \text{ and } D(x, y) \subseteq F.$$

Although this definition looks somewhat contrived, it is actually an old acquaintance in disguise. To recognize this, first notice that if $\mathcal{B}$ is normal over $\mathcal{R}$ then

$$x \equiv y(\mathcal{U}) \text{ iff } D(x, y) \notin \mathcal{U}$$

Next, if $\mathfrak{F}$ is a field of subsets and $\mathcal{B}$ is normal over $\mathfrak{F}$ then

$$x \equiv y(\mathcal{U}) \text{ iff } E(x, y) \in \mathcal{U}.$$

Finally, if $\mathcal{B} = \Pi\langle \mathfrak{A}_i \mid i \in I\rangle$ and $\mathfrak{F}$ is the power set of I then $\mathcal{B}$ is normal over $\mathfrak{F}$ and we obtain the congruence inducing the ultraproduct $\Pi_{\mathcal{U}}\langle \mathfrak{A}_i \mid i \in I\rangle$. Thus we actually have a natural generalization of the ultraproduct construction.

Lemma 4.7 If $\mathcal{U}$ is a prime filter on $\mathcal{R}$ then

(i) $\equiv(\mathcal{U})$ is a congruence on $\mathcal{B}$;

(ii) for every $R \in Rl$ and $x, y \in B^r$, if $x_p \equiv y_p(\mathcal{U})$ for every $p < r$, then there exists $F \in \mathcal{R}$ such that $F \notin \mathcal{U}$ and $[\![\neg R(x)]\!] \subseteq F$ iff there exists $F \in \mathcal{R}$ such that $F \notin \mathcal{U}$ and $[\![\neg R(y)]\!] \subseteq F$.

Proof: (i) We shall check transitivity. The remainder of the argument is similar. Suppose $F, G \in \mathcal{R}$, where $F, G \notin \mathcal{U}$ and $D(x, y) \subseteq F$, $D(y, z) \subseteq G$. Then $F \cup G \notin \mathcal{U}$ and

$$D(x, z) \subseteq D(x, y) \cup D(y, z) \subseteq F \cup G.$$

(ii) Assume for each $p < r$, $F_p \in \mathcal{R}$, $F_p \notin \mathcal{U}$ and $D(x_p, y_p) \subseteq F_p$. Consider $F \in \mathcal{R}$, where $F \notin \mathcal{U}$ and $[\![\neg R(x)]\!] \subseteq F$. Then we obtain $F \cup \bigcup_{p<r} F_p \notin \mathcal{U}$ and

$$[\![\neg R(y)]\!] \subseteq [\![\neg R(x)]\!] \cup \bigcup_{p<r} D(x_p, y_p) \subseteq F \cup \bigcup_{p<r} F_p$$

The assertion follows by symmetry. ∎

Lemma 4.7 permits us to define $\mathcal{L}/\mathcal{U}$, where

$R^{\mathcal{L}/\mathcal{U}}(x_o/\mathcal{U}, \ldots, x_{r-1}/\mathcal{U})$ iff for some $F \in \mathcal{R}$, $F \notin \mathcal{U}$ and $[\![\neg R(x)]\!] \subseteq F$.

For any $F \subseteq I$, F^o denotes the <u>interior</u> of F. We shall now use the notation introduced in Section 1 for the Wallman compactification.

<u>Lemma</u> 4.8 (i) For every $\mathcal{U} \in J$, every atomic formula φ and every $x \in B^n$,

$\mathcal{L}/\mathcal{U} \models \varphi[x/\mathcal{U}]$ iff for some $F \in \mathcal{R}$, $F \notin \mathcal{U}$ and $[\![\neg \varphi(x)]\!] \subseteq F$.

(ii) If $\mathcal{R}$ covers I then for every $i \in I$, every atomic formula φ and every $x \in B^n$,

$\mathcal{L}/h(i) \models \varphi[x/h(i)]$ iff $i \in [\![\varphi(x)]\!]^o$.

<u>Proof</u>: (i) is immediate. To prove (ii) use (i) and notice that there exists $F \in \mathcal{R}$ such that $i \notin F$ and $[\![\neg \varphi(x)]\!] \subseteq F$ if and only if there exists $F \in \mathcal{R}$ such that $i \in I - F \subseteq [\![\varphi(x)]\!]$. ∎

Now we define the <u>Gelfand</u> <u>mapping</u> to be the canonical mapping $g: B \to \Pi\langle B/\mathcal{U} \mid \mathcal{U} \in J\rangle$.

<u>Lemma</u> 4.9 (i) If $h(i) \neq \emptyset$ and $x \equiv y(h(i))$ then $x(i) = y(i)$.

(ii) If $h(i) \neq \emptyset$ and $R^{\mathcal{L}/h(i)}(x_o/h(i), \ldots, x_{r-1}/h(i))$ then $R^{\mathfrak{A}_i}(x_o(i), \ldots, x_{r-1}(i))$.

(iii) If $\mathcal{R}$ covers I then g is a subdirect embedding.

<u>Proof</u>: (i) and (ii) follow from Lemma 4.8 and (iii) follows from (i) and (ii). ∎

Lemma 4.10 For every atomic formula φ and every $x \in B^n$,

$$[\![\neg\varphi(g(x))]\!] = \bigcap\{h^*(F) \mid [\![\neg\varphi(x)]\!] \subseteq F\}.$$

Proof: $\mathcal{U} \in [\![\neg\varphi(g(x))]\!]$ if and only if $\mathcal{B}/\mathcal{U} \models \neg\varphi[x/\mathcal{U}]$. The assertion follows from Lemma 4.8. ∎

Corollary 4.11 If $\mathcal{R}$ covers I then for every atomic formula φ, every $x \in B^n$ and every $F \in \mathcal{R}$,

$$[\![\neg\varphi(x)]\!] \subseteq F \quad \text{iff} \quad [\![\neg\varphi(g(x))]\!] \subseteq h^*(F).$$

Proof: If $i \in [\![\neg\varphi(x)]\!]$ but $i \notin F$ then $h(i) \in [\![\neg\varphi(g(x))]\!]$ but $h(i) \notin h^*(F)$. The assertion follows from Lemma 4.10. ∎

Now recall from Section 1 that $\mathcal{T}(\mathcal{R}^*)$ is the topology on J with basis of closed sets

$$\mathcal{R}^* = \{h^*(F) \mid F \in \mathcal{R}\}.$$

Lemma 4.12 If $\mathcal{R}$ covers I then $g(\mathcal{B}) \subseteq \Pi\langle\mathcal{B}/\mathcal{U} \mid \mathcal{U} \in J\rangle$ is semi normal over $\mathcal{R}^*$.

Proof: Suppose $r < \omega$, $F_p \in \mathcal{R}$ and $x_p \in B$ for each $p < r$, where $\bigcap_{p<r} h^*(F_p) = \emptyset$ and $D(g(x_p), g(x_q)) \subseteq h^*(F_p) \cup h^*(F_q)$ whenever $p, q < r$. Then $h^*(\emptyset) = h^*(\bigcap_{p<r} F_p)$ and therefore $\bigcap_{p<r} F_p = \emptyset$. Moreover, $D(g(x_p), g(x_q)) \subseteq h^*(F_p \cup F_q)$ and therefore, by Corollary 4.11, $D(x_p, x_q) \subseteq F_p \cup F_q$ whenever $p, q < r$. Since $\mathcal{B}$ patches over $\breve{\mathcal{R}}$, by Lemma 4.2 there exists $y \in B$ such that $D(y, x_p) \subseteq F_p$ for all $p < r$. By Corollary 4.11, $D(g(y), g(x_p)) \subseteq h^*(F_p)$ for all $p < r$ and using Lemma 4.2 again we have shown that $g(\mathcal{B})$ patches over $\breve{\mathcal{R}}^*$. The assertion follows from Lemma 4.10. ∎

We can now state the central result of this study:

Theorem 4.13 Suppose $\mathcal{B} \subseteq \Pi\langle\mathfrak{A}_i \mid i \in I\rangle$ is subdirect and $\mathcal{R}$ is a ring of subsets of I covering I. If $\mathcal{B}$ patches over $\breve{\mathcal{R}}$ then $\mathcal{B} \cong g(\mathcal{B}) \subseteq \Pi\langle\mathcal{B}/\mathcal{U} \mid \mathcal{U} \in J\rangle$ is $\mathcal{T}(\mathcal{R}^*)$-global.

Proof: Use Lemma 1.1, 4.4, 4.9(iii) and 4.12. ∎

Corollary 4.14 Every subdirect product which patches over a dual ring $\breve{\mathcal{R}}$ of subsets of the index set, where $\bigcap \mathcal{R} = \emptyset$, is isomorphic to a global subdirect product over a compact T_o-space with an open basis isomorphic to $\breve{\mathcal{R}}$. ∎

As far as we have been able to ascertain, Theorem 4.13 covers every (discrete) sheaf representation result in the literature starting from finite patching. Three issues remain to be investigated: Where does the ring $\mathcal{R}$ of subsets come from? How is the patching accomplished? What are the new factors of the global subdirect representation? We shall address ourselves to the last issue first. For this purpose, again let $\mathcal{B} \subseteq \Pi\langle \mathfrak{A}_i \mid i \in I\rangle$ be subdirect and let $\mathcal{R}$ be a ring of subsets of I covering I. Assume that $\mathcal{B}$ patches over $\breve{\mathcal{R}}$. By Lemma 4.9, for each $i \in I$ the canonical mapping $g_i: B/h(i) \to A_i$ is a surjective homomorphism.

Corollary 4.15 Suppose $\mathcal{B} \subseteq \Pi\langle \mathfrak{A}_i \mid i \in I\rangle$ is subdirect and $\mathcal{R}$ is a ring of subsets of I covering I. If $\mathcal{B}$ patches over $\breve{\mathcal{R}}$ then $\mathcal{B}$ is semi normal over $\mathcal{R}$ if and only if for every $i \in I$, $g_i: \mathcal{B}/h(i) \to \mathfrak{A}_i$ is an isomorphism.

Proof: Suppose for every $i \in I$, g_i is an isomorphism. Let φ be an atomic formula and let $x \in B^n$. If $i \in [\![\varphi(x)]\!]$ then $\mathfrak{A}_i \models \varphi[\pi_i(x)]$ and therefore, by hypothesis, $\mathcal{B}/h(i) \models \varphi[x/h(i)]$. Thus by Lemma 4.8, $i \in [\![\varphi(x)]\!]^o$ and $\mathcal{B}$ is semi normal over $\mathcal{R}$. Conversely, suppose $\mathcal{B}$ is semi normal over $\mathcal{R}$. Let φ be an atomic formula and let $x \in B^n$. If $\mathfrak{A}_i \models \varphi[\pi_i(x)]$ then $i \in [\![\varphi(x)]\!]$ and therefore, by hypothesis, $i \in [\![\varphi(x)]\!]^o$. By Lemma 4.8, $\mathcal{B}/h(i) \models \varphi[x/h(i)]$ and it follows that g_i is an isomorphism. ∎

Globalizing a subdirect product $\mathcal{B} \subseteq \Pi\langle \mathfrak{A}_i \mid i \in I\rangle$ which patches over a dual ring $\breve{\mathcal{R}}$ of subsets of I the old factors $\mathfrak{A}_i$ are "blown up" into new factors $\mathcal{B}/h(i)$, i.e. the old factors $\mathfrak{A}_i$ are homomorphic images of the new factors $\mathcal{B}/h(i)$. Corollary 4.15 tells us that we get all old factors back if and only if $\mathcal{B}$ is semi normal over $\mathcal{R}$. Next we shall look at the new factors which are not "blown up" old factors. If $\mathcal{R}$ covers I and $h(I) \subseteq K \subseteq J$ then by Lemma 4.8, the projection $\pi_K: g(\mathcal{B}) \to \Pi\langle \mathcal{B}/\mathcal{U} \mid \mathcal{U} \in K\rangle$ is an isomorphism. To simplify the notation, let $g_K = \pi_K \circ g$ so that we have the following commutative diagram:

$$\begin{array}{ccl} & \overset{g}{\nearrow} & \Pi\langle \mathcal{B}/\mathcal{U} \mid \mathcal{U} \in J\rangle \\ \mathcal{B} & & \quad \downarrow \pi_K \\ & \underset{g_K}{\searrow} & \Pi\langle \mathcal{B}/\mathcal{U} \mid \mathcal{U} \in K\rangle \end{array}$$

Lemma 4.16 Suppose $\mathcal{R}$ covers I and $h(I) \subseteq K \subseteq J$. Then for every atomic formula φ, every $x \in B^n$ and every $F \in \mathcal{R}$,

$$[\![\neg\varphi(x)]\!] \subseteq F \quad \text{iff} \quad [\![\neg\varphi(g(x))]\!] \cap K \subseteq h^*(F) \cap K.$$

Proof: Consult the proof of Corollary 4.11. ∎

Corollary 4.17 Suppose $\mathcal{R}$ covers I and $h(I) \subseteq K \subseteq J$. Then $g_K(\mathcal{B}) \subseteq \Pi\langle \mathcal{B}/\mathcal{U} \mid \mathcal{U} \in K\rangle$ is semi normal over $\mathcal{R}^* \mid K = \{h^*(F) \cap K \mid F \in \mathcal{R}\}$.

Proof: By Lemmas 2.36 and 4.10, for every atomic formula φ and every $x \in B^n$, $[\![\neg\varphi(g_K(x))]\!]$ is $\mathfrak{T}(\mathcal{R}^* \mid K)$-closed. Now use Lemma 4.16 and argue as in the proof of Lemma 4.12. ∎

Now we obtain a generalization of Theoren 4.13.

Theorem 4.18 Suppose $\mathcal{B} \subseteq \Pi\langle \mathfrak{A}_i \mid i \in I\rangle$ is subdirect and $\mathcal{R}$ is a ring of subsets of I covering I. If $\mathcal{B}$ patches over $\breve{\mathcal{R}}$ and $h(I) \subseteq K \subseteq J$, where K is compact, then $\mathcal{B} \cong g_K(\mathcal{B}) \subseteq \Pi\langle \mathcal{B}/\mathcal{U} \mid \mathcal{U} \in K\rangle$ is $\mathfrak{T}(\mathcal{R}^* \mid K)$-global.

Proof: Use Lemma 4.4 and Corollary 4.17. ∎

Of course, it is desirable to "obtain K as small as possible".

Corollary 4.19 Suppose $\mathcal{B} \subseteq \Pi\langle \mathfrak{A}_i \mid i \in I\rangle$ is subdirect and $\mathcal{R}$ is a ring of subsets of I covering I. If $\mathcal{B}$ patches over $\breve{\mathcal{R}}$ and I is compact then $\mathcal{B} \cong g_{h(I)}(\mathcal{B}) \subseteq \Pi\langle \mathcal{B}/\mathcal{U} \mid \mathcal{U} \in h(I)\rangle$ is $\mathfrak{T}(\mathcal{R}^* \mid h(I))$-global.

Proof: If I is compact then $h(I)$ is compact. ∎

Remarks 4.20 Corollary 4.19 has several important applications in the literature on sheaf representation. In this special case the globalization of $\mathcal{B}$ can be obtained somewhat more directly by keeping the index space I. Indeed, with the canonical embedding

$$g_I : \mathcal{B} \to \Pi\langle \mathcal{B}/h(i) \mid i \in I\rangle$$

$g_I(\mathcal{B}) \subseteq \Pi\langle \mathcal{B}/h(i) \mid i \in I\rangle$ is semi normal over $\mathcal{R}$ in case $\mathcal{B}$ patches over $\breve{\mathcal{R}}$. Hence, if I is compact then $\mathcal{B} \cong g_I(\mathcal{B}) \subseteq \Pi\langle \mathcal{B}/h(i) \mid i \in I\rangle$ is $\mathfrak{I}(\mathcal{R})$-global. If I is a T_o-space then by Lemma 1.4, h is onte-to-one and the two constructions are the same. This is actually the case in most applications.

Corollary 4.21 Suppose $\mathcal{B} \subseteq \Pi\langle \mathfrak{A}_i \mid i \in I\rangle$ is subdirect and $\mathcal{R}$ is a ring of subsets of I covering I. If $\mathcal{B}$ patches over $\breve{\mathcal{R}}$ and every open subset of I is the union of closed sets then $\mathcal{B} \cong g_M(\mathcal{B}) \subseteq \Pi\langle \mathcal{B}/\mathcal{U} \mid \mathcal{U} \in M\rangle$ is $\mathfrak{I}(\mathcal{R}^* \mid M)$-global, where M is the set of maximal filters on $\mathcal{R}$.

Proof: By Lemma 1.7, $h(I) \subseteq M \subseteq J$ and by Lemma 1.9, M is compact. ∎

Notice, if $\mathcal{R}$ is a Boolean ring (or, a fortiori, a field) of subsets of I then every prime filter on $\mathcal{R}$ is maximal so that in this case Theorem 4.13 and Corollary 4.21 coincide. In general there is not much we can say about the factors $\mathcal{B}/\mathcal{U}$ unless $\mathcal{U}$ is a maximal filter and $\mathcal{B}$ is normal over $\mathcal{R}$. These are severe restrictions which exclude many examples from the literature. Maybe this explains why in these examples the authors give no account of the factors whatsoever.

Lemma 4.22 Suppose $\mathcal{B} \subseteq \Pi\langle \mathfrak{A}_i \mid i \in I\rangle$ is subdirect and $\mathcal{R}$ is a ring of subsets of I. If $\mathcal{B}$ is normal over $\mathcal{R}$ then

(i) for every prime filter $\mathcal{U}$ on $\mathcal{R}$, every atomic formula φ and every $x \in B^n$,

$$\mathcal{B}/\mathcal{U} \models \neg\varphi[x/\mathcal{U}] \text{ iff } [\![\neg\varphi(x)]\!] \in \mathcal{U};$$

(ii) for every atomic formula φ and every $x \in B^n$,

$$h^*([\![\neg\varphi(x)]\!]) = [\![\neg\varphi(g(x))]\!]. \blacksquare$$

The next result is a rather simple obersevation. Nevertheless, it provides the best information available in general concerning the new factors.

Theorem 4.23 Suppose $\mathcal{B} \subseteq \Pi\langle \mathfrak{A}_i \mid i \in I\rangle$ is subdirect and $\mathcal{R}$ is a ring of subsets of I covering I, where $\mathcal{B}$ is normal over $\mathcal{R}$.

(i) If $\mathcal{U}$ is a prime filter on $\mathcal{R}$ then $\mathcal{B}/\mathcal{U} \in HSP_u\{\mathfrak{A}_i \mid i \in I\}$.

(ii) If $\mathcal{U}$ is a maximal filter on $\mathcal{R}$ then $\mathcal{B}/\mathcal{U} \in ISP_u\{\mathfrak{A}_i \mid i \in I\}$.

Proof: (i) Let $\mathcal{V}$ be an ultrafilter on I such that $\mathcal{U} \subseteq \mathcal{V}$. If φ is an atomic formula and $x \in B^n$ where not $\mathcal{B}/\mathcal{U} \models \varphi[x/\mathcal{U}]$ then by

Lemma 4.22, $[\![\neg\varphi(x)]\!] \in \mathcal{U}$ and therefore $[\![\neg\varphi(x)]\!] \in \mathcal{V}$. Thus not $\Pi_{\mathcal{V}}\langle \mathfrak{A}_i \mid i \in I\rangle \models \varphi[x/\mathcal{V}]$ and it follows by the canonical construction that $\mathcal{B}/\mathcal{U} \in HSP_u\{\mathfrak{A}_i \mid i \in I\}$.

(ii) If $\mathcal{U}$ is a maximal filter on $\mathcal{R}$ then for any atomic formula φ and any $x \in B^n$, $[\![\neg\varphi(x)]\!] \in \mathcal{U}$ if and only if $[\![\neg\varphi(x)]\!] \in \mathcal{V}$. It follows that $\mathcal{B}/\mathcal{U} \in ISP_u\{\mathfrak{A}_i \mid i \in I\}$. ∎

Corollary 4.21 tells us when we get by with maximal filters on $\mathcal{R}$, and Theorem 4.23 gives us information about the new factors which come from maximal filters on $\mathcal{R}$. We shall see later that the process of compactification of the index set frequently makes the appearance of trivial new factors unavoidable. The next lemma tells us when all new factors which come from maximal filters are either trivial or are old factors.

<u>Lemma</u> 4.24 Suppose $\mathcal{B} \subseteq \Pi\langle \mathfrak{A}_i \mid i \in I\rangle$ is subdirect and $\mathcal{R}$ is a ring of subsets of I, where $\mathcal{B}$ is normal over $\mathcal{R}$. Then the following are equivalent:

(i) For every maximal filter $\mathcal{U}$ on $\mathcal{R}$, either $\mathcal{B}/\mathcal{U}$ is trivial or there exists $i \in I$ such that $\mathcal{U} = h(i)$.

(ii) For every $x, y \in B$, $D(x, y)$ is compact.

<u>Proof</u>: Assume (i) and let $x, y \in B$. Suppose $F_\xi \in \mathcal{R}$ for each $\xi \in \Delta$, where $\{F_\xi \cap D(x, y) \mid \xi \in \Delta\}$ has the finite intersection property. Then there exists a maximal filter $\mathcal{U}$ on $\mathcal{R}$ such that $D(x, y) \in \mathcal{U}$ and $F_\xi \in \mathcal{U}$ for all $\xi \in \Delta$. Since $\mathcal{B}/\mathcal{U}$ is non-trivial, there exists $i \in I$ such that $\mathcal{U} = h(i)$ and therefore

$$D(x, y) \cap \bigcap_{\xi\in\Delta} F_\xi = \emptyset.$$

Conversely, assume (ii) and let $\mathcal{U}$ be a maximal filter on $\mathcal{R}$, where $\mathcal{B}/\mathcal{U}$ is non-trivial. Then there exist $x, y \in B$ such that $D(x, y) \in \mathcal{U}$ and therefore $\{F \cap D(x, y) \mid F \in \mathcal{U}\}$ has the finite intersection property. By hypothesis there exists $i \in \bigcap\mathcal{U}$. Thus $\mathcal{U} \subseteq h(i)$ and since $\mathcal{U}$ is maximal, $\mathcal{U} = h(i)$. ∎

Finally we shall collect all the information about the factors in the most favorable case.

<u>Theorem</u> 4.25 Suppose $\mathcal{B} \subseteq \Pi\langle \mathfrak{A}_i \mid i \in I\rangle$ is subdirect and $\mathcal{R}$ is a ring of subsets of I covering I. If $\mathcal{B}$ is normal over $\mathcal{R}$ and every open

subset of I is the union of closed sets then $\mathcal{B} \cong g_M(\mathcal{B}) \subseteq \Pi\langle \mathcal{B}/\mathcal{U} \mid \mathcal{U} \in M\rangle$ is $\mathcal{T}(\mathcal{R}^* \upharpoonright M)$-global, where M is the set of maximal filters on $\mathcal{R}$. Moreover,

(i) for every $i \in I$, $\mathcal{B}/h(i) \cong \mathfrak{A}_i$;

(ii) for every $\mathcal{U} \in M$, $\mathcal{B}/\mathcal{U} \in O\{\mathfrak{A}_i \mid i \in I\}$;

(iii) if $D(x, y)$ is compact for every $x, y \in B$ then for every $\mathcal{U} \in M$, $\mathcal{B}/\mathcal{U}$ is trivial or there exists $i \in I$ such that $\mathcal{U} = h(i)$. ∎

There is a way of skirting the issue of the new factors by starting with enough old factors to insure that the equalizer topology is compact. We give the construction which is due to Kennison [26].

<u>Lemma</u> 4.26 Suppose an algebra $\mathcal{B} \subseteq \Pi\langle \mathfrak{A}_i \mid i \in I\rangle$ is subdirect and $\mathcal{M}$ is a universal class, where $\{\theta_i^{\mathcal{B}} \mid i \in I\} = \{\psi \mid \mathcal{B}/\psi \in \mathcal{M}\}$. Let $\mathcal{T}$ be the smallest topology on I such that for all $x, y \in B$, $E(x, y)$, $D(x, y) \in \mathcal{T}$. Then I (and, a fortiori, with the equalizer topology induced by $\mathcal{B}$) is compact.

<u>Proof</u>: Let $\mathcal{U}$ be an ultrafilter on I. Since $\Pi_{\mathcal{U}}\langle \mathfrak{A}_i \mid i \in I\rangle \in \mathcal{M}$ there exists $j \in I$ such that for all $x, y \in B$, $\pi_j(x) = \pi_j(y)$ if and only if $E(x, y) \in \mathcal{U}$. Thus $j \in E(x, y)$ if and only if $E(x, y) \in \mathcal{U}$, so that $\mathcal{U}$ converges to j. ∎

<u>Corollary</u> 4.27 Suppose an algebra $\mathcal{B} \subseteq \Pi\langle \mathfrak{A}_i \mid i \in I\rangle$ is subdirect and $\mathcal{M}$ is a universal class, where $\{\theta_i^{\mathcal{B}} \mid i \in I\} = \{\psi \mid \mathcal{B}/\psi \in \mathcal{M}\}$. Then $\mathcal{B}$ is global if and only if $\mathcal{B}$ patches over the dual ring generated by $\{E(x, y) \mid x, y \in B\}$.

<u>Proof</u>: Use Lemmas 4.4 and 4.26. ∎

Next we shall comment on the problem of triviality of the globalization. For the sake of simplicity we shall only consider <u>algebras</u>. We should also emphasize again that some authors in sheaf representation appear to have no idea whatsoever whether their representation is non-trivial.

<u>Lemma</u> 4.28 Suppose an algebra $\mathcal{B} \subseteq \Pi\langle \mathfrak{A}_i \mid i \in I\rangle$ is subdirect and $\mathcal{R}$ is a ring of subsets of I covering I, where $\mathcal{B}$ is normal over $\mathcal{R}$. Then $g(\mathcal{B}) \subseteq \Pi\langle \mathcal{B}/\mathcal{U} \mid \mathcal{U} \in J\rangle$ is non-trivially subdirect if and only if

$$\{D(x, y) \mid x, y \in B, x \neq y\}$$

does not have the finite intersection property.

Proof: Use Lemma 4.22. ∎

Corollary 4.29 Suppose an algebra $\mathcal{B} \subseteq \Pi\langle \mathfrak{A}_i \mid i \in I\rangle$ is subdirect and $\mathcal{R}$ is a ring of subsets of I covering I, where $\mathcal{B}$ is normal over $\mathcal{R}$. If there exist $x, y, z \in B$ such that $E(x, y) \in \mathcal{R}$ and $E(x, y) \cap D(x, z) \neq \emptyset$ then $g(\mathcal{B}) \subseteq \Pi\langle \mathcal{B}/\mathcal{U} \mid \mathcal{U} \in J\rangle$ is non-trivially subdirect.

Proof: Since $\mathcal{B}$ is normal over $\mathcal{R}$, $\overline{t}(x, y, z) \in B$. By hypothesis, $\overline{t}(x, y, z) \neq x$ and $D(x, y) \cap D(\overline{t}(x, y, z), x) = \emptyset$. The assertion follows from Lemma 4.28. ∎

Next we shall pursue the second question raised after Corollary 4.14: How is the patching accomplished? Although some patching arguments can be carried out in the setting of universal algebra, here frequently some special information from algebra is required. In fact, this is usually where the non-trivial part of sheaf constructions takes place. The question is: Does a given structure $\mathcal{B}$ have a subdirect representation which patches over some (non-trivial) dual-ring of subsets of the index set? The natural place to look for an answer is the congruence lattice of $\mathcal{B}$, and in the case of patching over a field of subsets the answer is given by the Comer-Pierce construction (Comer [8], Pierce [31]).

Lemma 4.30 Suppose $\mathcal{B} \subseteq \Pi\langle \mathfrak{A}_i \mid i \in I\rangle$ and $\mathcal{F}$ is a field of subsets of I. Then the following are equivalent:

(i) $\mathcal{B}$ patches over $\mathcal{F}$.

(ii) For every $F, G \in \mathcal{F}$,

$$\theta^{\mathcal{B}}_{F\cap G} = \theta^{\mathcal{B}}_F \vee \theta^{\mathcal{B}}_G = \theta^{\mathcal{B}}_F \theta^{\mathcal{B}}_G$$

(iii) The mapping $F \in \mathcal{F} \to \theta^{\mathcal{B}}_F$ is an anti homomorphism onto a sublattice of the congruence lattice of $\mathcal{B}$ containing $\mathbb{O}^{\mathcal{B}}$ and $\mathbb{1}^{\mathcal{B}}$, which is a Boolean algebra of mutually permuting congruences.

Proof: The equivalence of (ii) and (iii) follows from Lemma 3.15. Assume (i) and let $F, G \in \mathcal{F}$. Since always

$$\theta^{\mathcal{B}}_F \theta^{\mathcal{B}}_G \subseteq \theta^{\mathcal{B}}_F \vee \theta^{\mathcal{B}}_G \subseteq \theta^{\mathcal{B}}_{F\cap G},$$

to prove (ii) it suffices to show that $\theta^{\mathcal{B}}_{F\cap G} \subseteq \theta^{\mathcal{B}}_F \theta^{\mathcal{B}}_G$. Suppose $F \cap G \subseteq E(x, y)$. Let

$$z(i) = \begin{cases} x(i) & \text{if } i \in F \\ y(i) & \text{if } i \notin F \end{cases}$$

Then $z \in B$ and $F \subseteq E(x, z)$, $G \subseteq E(z, y)$. This establishes (ii).

Conversely, assume (ii) and let $F \in \mathfrak{F}$, $x, y \in B$ and

$$z(i) = \begin{cases} x(i) & \text{if } i \in F \\ y(i) & \text{if } i \notin F. \end{cases}$$

Since

$$\mathbb{1}^{\mathcal{L}} = \theta^{\mathcal{L}}_{F\cap(I-F)} = \theta^{\mathcal{L}}_{F}\,\theta^{\mathcal{L}}_{I-F}$$

there exists $w \in B$ such that $x\theta^{\mathcal{L}}_{F}w$ and $w\theta^{\mathcal{L}}_{I-F}y$. Thus $w = z$ and by Lemma 4.5, (i) is established. ∎

Lemma 4.30 reveals that patching over a field induces a sublattice of the congruence lattice of $\mathcal{L}$ containing $\mathbb{0}^{\mathcal{L}}$ and $\mathbb{1}^{\mathcal{L}}$, which is a Boolean algebra of permuting congruences. Comer [8] uses the existence of such a sublattice of the congruence lattice to give a global subdirect representation for $\mathcal{L}$. We quickly give the construction for an <u>algebra</u> $\mathcal{L}$. Let $\mathfrak{B}$ be a sublattice of the congruence lattice of $\mathcal{L}$ containing $\mathbb{0}^{\mathcal{L}}$ and $\mathbb{1}^{\mathcal{L}}$, which is a Boolean algebra of permuting congruences. Let J be the set of prime filters of $\mathfrak{B}$. For each $\mathcal{U} \in J$, let

$$\hat{\mathcal{U}} = \bigcup\{\theta \in \mathfrak{B} \mid \theta \notin \mathcal{U}\}.$$

Then $\hat{\mathcal{U}}$ is a congruence on $\mathfrak{B}$.

<u>Lemma</u> 4.31 $\bigcap\{\hat{\mathcal{U}} \mid \mathcal{U} \in J\} = \mathbb{0}^{\mathcal{L}}$.

<u>Proof</u>: Let $x, y \in B$, where $x \neq y$. Then $\{\theta \in \mathfrak{B} \mid x\theta y\}$ is a proper filter on $\mathfrak{B}$ which is contained in a maximal filter $\mathcal{U}$ on $\mathfrak{B}$. Clearly $(x, y) \notin \hat{\mathcal{U}}$. ∎

Now subdirectly embed

$$g: \mathcal{L} \to \Pi\langle \mathcal{L}/\hat{\mathcal{U}} \mid \mathcal{U} \in J\rangle.$$

J is the dual Stone space of $\mathfrak{B}$ with basis of <u>clopen</u> sets

$$h^*(\theta) = \{\mathcal{U} \in J \mid \theta \notin \mathcal{U}\}$$

where h^* is an anti isomorphism from $\mathfrak{B}$ onto the basis of clopen sets.

<u>Lemma</u> 4.32 For every $x, y \in B$,

$$E(g(x), g(y)) = \bigcup\{h^*(\theta) \mid x\theta y\}.$$

Proof: $\mathcal{U} \in E(g(x), g(y))$ if and only if $x/\hat{\mathcal{U}} = y/\hat{\mathcal{U}}$, if and only if for some $\theta \in \mathcal{B}$, $\mathcal{U} \in h^*(\theta)$ and $x\theta y$. ∎

Lemma 4.33 For all $\chi \in \mathcal{B}$,

$$x\chi y \text{ iff } g(x)\theta^{g(\mathcal{L})}_{h^*(\chi)}g(y).$$

Proof: Suppose $x\chi y$. Then for every $\mathcal{U} \in h^*(\chi)$, $\langle x, y\rangle \in \hat{\mathcal{U}}$. Thus $g(x)\theta^{g(\mathcal{L})}_{h^*(\chi)}g(y)$. Conversely, suppose not $x\chi y$. Then $\{\theta \in \mathcal{B} \mid x\theta y\}$ is a proper filter on $\mathcal{B}$. Let $\mathcal{U}$ be a filter on $\mathcal{B}$ which is maximal with respect to $\{\theta \in \mathcal{B} \mid x\theta y\} \subseteq \mathcal{U}$ and $\chi \notin \mathcal{U}$. Then $\mathcal{U} \in J$, $\mathcal{U} \in h^*(\chi)$ and $\langle x, y\rangle \notin \hat{\mathcal{U}}$. Thus not $g(x)\theta^{g(\mathcal{L})}_{h^*(\chi)} g(y)$. ∎

Theorem 4.34 (Comer) Let $\mathcal{B}$ be a sublattice of the congruence lattice of an algebra $\mathcal{L}$ containing $\mathbb{0}^{\mathcal{L}}$ and $\mathbb{1}^{\mathcal{L}}$, which is a Boolean algebra of permuting congruences, and let J be the set of prime filters of $\mathcal{B}$. Then $\mathcal{L} \cong g(\mathcal{L}) \subseteq \Pi\langle \mathcal{L}/\hat{\mathcal{U}} \mid \mathcal{U} \in J\rangle$ is $T(\mathcal{B}^*)$-global.

Proof: We claim that for each $\varphi, \psi \in \mathcal{B}$

$$\theta^{g(\mathcal{L})}_{h^*(\varphi)\cap h^*(\psi)} \subseteq \theta^{g(\mathcal{L})}_{h^*(\varphi)}\theta^{g(\mathcal{L})}_{h^*(\psi)} .$$

Indeed, suppose $g(x)\theta^{g(\mathcal{L})}_{h^*(\varphi)\cap h^*(\psi)} g(y)$. Then $g(x)\theta^{g(\mathcal{L})}_{h^*(\varphi\vee\psi)}g(y)$ and therefore, by Lemma 4.33, $x(\varphi \vee \psi)y$. By hypothesis, there exists $z \in B$ such that $x\varphi z\psi y$ and by Lemma 4.33, $g(x)\theta^{g(\mathcal{L})}_{h^*(\varphi)}g(z)\theta^{g(\mathcal{L})}_{h^*(\psi)}g(y)$. This establishes our claim. By Lemmas 4.30 and 4.32, $g(\mathcal{L})$ is semi normal over $\mathcal{B}^* = \{h^*(\theta) \mid \theta \in \mathcal{B}\}$, and the assertion follows from Lemma 4.4. ∎

It is natural to ask for an analogue of Lemma 4.30 for patching over a dual ring. Unfortunately there does not appear to be such an analogue unless the notion of patching is strengthened. Let $\mathcal{L} \subseteq \Pi\langle \mathfrak{A}_i \mid i \in I\rangle$ and let $\mathcal{R}$ be a dual ring of subsets of I. We say that $\mathcal{L}$ strongly patches over $\mathcal{R}$ if for every $r < \omega$, if $F_p \in \mathcal{R}$ and $x_p \in B$ for each $p < r$, where $F_p \cap F_q \subseteq E(x_p, x_q)$ whenever $p, q < r$, then there exists $y \in B$ such that $F_p \subseteq E(y, x_p)$ for all $p < r$. Of course, if $\mathcal{L}$ strongly patches over $\mathcal{R}$ then $\mathcal{L}$ patches over $\mathcal{R}$, however the converse is not true in general. In fact, if $\mathcal{B}$ is $\mathcal{T}$-global then $\mathcal{B}$ does not in general strongly patch over the dual ring of open sets (Compare Lemma 4.3).

Lemma 4.35 Suppose $\mathcal{B} \subseteq \Pi\langle \mathfrak{A}_i \mid i \in I\rangle$ and $\mathcal{R}$ is a dual ring of subsets of I. The the following are equivalent:

(i) $\mathcal{B}$ strongly patches over $\mathcal{R}$.

(ii) For every $F, G \in \mathcal{R}$,

$$\theta^{\mathcal{B}}_{F\cap G} = \theta^{\mathcal{B}}_F \vee \theta^{\mathcal{B}}_G = \theta^{\mathcal{B}}_F \theta^{\mathcal{B}}_G.$$

(iii) The mapping $F \in \mathcal{R} \to \theta^{\mathcal{B}}_F$ is an anti homomorphism onto a sublattice of the congruence lattice of $\mathcal{B}$ containing $\mathcal{O}^{\mathcal{B}}$, which is a distributive lattice of mutually permuting congruences.

Proof: The equivalence of (ii) and (iii) again follows from Lemma 3.15. Assume (i) and let $F, G \in \mathcal{R}$, where $F \cap G \subseteq E(x, y)$. Then there exists $z \in B$ such that $F \subseteq E(z, x)$ and $G \subseteq E(z, y)$. This establishes that $\theta^{\mathcal{B}}_{F\cap G} \subseteq \theta^{\mathcal{B}}_F \theta^{\mathcal{B}}_G$ and (ii) follows.

Assume (ii) and let $r < \omega$ and $F_p \in \mathcal{R}$, $x_p \in B$ for each $p < r$, where $F_p \cap F_q \subseteq E(x_p, x_q)$ whenever $p, q < r$. Then for each $p, q < r$,

$$x_p \theta^{\mathcal{B}}_{F_p \cap F_q} x_q$$

and therefore

$$x_p \theta^{\mathcal{B}}_{F_p} \theta^{\mathcal{B}}_{F_q} x_q$$

By the Chinese Remainder Theorem there exists $y \in B$ such that $y \theta^{\mathcal{B}}_{F_p} x_p$ for all $p < r$ and therefore $F_p \subseteq E(y, x_p)$ for all $p < r$. ∎

Corollary 4.36 Suppose $\mathcal{B} \subseteq \Pi\langle \mathfrak{A}_i \mid i \in I\rangle$ and $\mathcal{R}$ is a dual ring of subsets of I. Then $\mathcal{B}$ strongly patches over $\mathcal{R}$ if and only if for every $F, G \in \mathcal{R}$ and every $x, y \in B$, where $F \cap G \subseteq E(x, y)$, there exists $z \in B$ such that $F \subseteq E(z, x)$ and $G \subseteq E(z, y)$.

Proof: Proving (i) implies (ii) in Lemma 4.35 we only need strong patching over two members of $\mathcal{R}$. ∎

Corollary 4.37 Suppose $\mathcal{B} \subseteq \Pi\langle \mathfrak{A}_i \mid i \in I\rangle$ and $\mathfrak{F}$ is a field of subsets of I. Then $\mathcal{B}$ patches over $\mathfrak{F}$ if and only if $\mathcal{B}$ strongly patches over $\mathfrak{F}$.

Proof: Use Lemmas 4.30 and 4.35. ∎

Lemma 4.35 reveals that strong patching over a ring induces a sublattice of the congruence lattice of $\mathcal{B}$ containing $\mathcal{D}^{\mathcal{B}}$ which is a distributive lattice of mutually permuting congruences. Wolf [41] uses the existence of such a sublattice of the congruence lattice to give a global subdirect representation for $\mathcal{B}$. We shall give an improved version of this result which is due to Werner [40]. The essence of Werner's improvement is the proof of Comer's Theorem which we have given above.

Theorem 4.38 (*Werner*) Let $\mathcal{L}$ be a sublattice of the congruence lattice of an algebra $\mathcal{B}$ containing $\mathcal{D}^{\mathcal{B}}$, which is a distributive lattice of permuting congruences, and let J be the set of prime filters of $\mathcal{L}$. Then $\mathcal{B} \cong g(\mathcal{B}) \subseteq \Pi\langle \mathcal{B}/\hat{\mathcal{U}} \mid \mathcal{U} \in J\rangle$ is $\mathfrak{T}(\mathcal{L}^*)$-global.

Proof: Repeat the proof of Theorem 4.34, where Lemma 4.30 is replaced by Lemma 4.35. ∎

Lemma 4.35 reveals that *strong patching over a ring* (a fortiori *patching over a field*) essentially comes from congruence permutability. However, *patching over a ring* does *not* always come from congruence permutability, and this is where the more subtle patching arguments can be found. We shall give several interesting examples in Sections 7 and 8.

Finally we shall address ourselves to the first issue raised after Corollary 4.14: *Where does the ring* $\mathcal{R}$ *of subsets come from*? Most applications start from a subdirect representation of an *algebra* $\mathcal{B} \subseteq \Pi\langle \mathcal{A}_i \mid i \in I\rangle$ inducing a hull-kernel topology on I and $\mathcal{R}$ is the *ring of all hull-kernel closed sets*. Notice that in this case $\mathcal{R}$ covers I. Nevertheless, it appears to us that in many cases the ring of hull-kernel closed sets is actually rather ill suited for globalization, and this accounts for the more peculiar features of the sheaf construction. The reason is obvious: Global subdirect products require a topology containing the *equalizer topology* on I induced by $\mathcal{B}$, which has subbasis

$$\{E(x, y) \mid x, y \in B\},$$

whereas the *hull-kernel topology* on I induced by $\mathcal{B}$ has basis

$$\{D(x, y) \mid x, y \in B\}.$$

Usually it is shown that $\mathcal{B}$ patches over the dual ring $\check{\mathcal{R}}$ of hull-kernel open sets so that Theorem 4.13 applies. However, in the majority of cases we have looked at, absolutely nothing is known about the new factors. In a few cases the hull-kernel topology is shown to be compact so that Corollary 4.19 (or Remarks 4.20) applies. However, even in these special cases usually very little can be said about the process of "blowing up" the factors. By Corollary 4.15, we only get the old factors back if $\mathcal{B}$ is semi normal over $\mathcal{R}$.

Lemma 4.39 Suppose an algebra $\mathcal{B} \subseteq \Pi\langle \mathfrak{A}_i \mid i \in I\rangle$ is subdirect and induces a hull-kernel topology on I. Let $\mathcal{R}$ be the ring of hull-kernel closed sets and assume that $\mathcal{B}$ patches over $\check{\mathcal{R}}$. Then the following are equivalent:

(i) $\mathcal{B}$ is semi normal over $\mathcal{R}$.

(ii) $\mathcal{B}$ is normal over $\mathcal{R}$.

(iii) For every $x, y \in B$, $E(x, y)$ is clopen.

(iv) $\mathcal{B}$ is normal over the field of clopen sets.

Proof: Use Lemma 3.1. ∎

Thus working with the hull-kernel closed sets the requirement of retaining the old factors already eradicates all subtler distinctions: equalizers are clopen! The rare cases where this actually occurs lead to the more interesting sheaf representations and will be further investigated in Section 5. In most other cases all we have been able to ascertain is that the "blown up" factors appear to be huge and only loosely connected with the old factors. A further "tightening" of the subdirect representation has been traded for our knowledge of the factors.

Of course, the natural thing to do is to consider the ring $\mathcal{R}$ of hull-kernel open sets. Now equalizers behave right, but the topology is turned around because $\mathcal{T}(\mathcal{R})$ is the topology with basis (of open sets) the dual ring $\check{\mathcal{R}}$ of hull-kernel closed sets. Although this construction appears to be somewhat peculiar, it does have interesting applications in case $\mathcal{B}$ is a congruence distributive algebra. By Theorem 3.24, we may assume that $\mathcal{B} \subseteq \Pi\langle \mathfrak{A}_i \mid i \in I\rangle$ is subdirect, where $\{\theta_i^{\mathcal{B}} \mid i \in I\}$ is a set of irreducible congruences containing all completely meet irreducible congruences. By Lemma 3.20, $\mathcal{B}$ is fully expanded and by Corollary 3.8, $\mathcal{B}$ induces a hull-kernel topology on I. The crucial issue is patching. By Corollary 3.17, the mapping $F \to \theta_F^{\mathcal{B}}$ is an anti isomorphism between the ring of hull-kernel closed sets and the congruence lattice of $\mathcal{B}$.

Lemma 4.40 Let $\mathcal{B}$ be a congruence distributive algebra and let $\mathcal{B} \subseteq \Pi\langle \mathfrak{A}_i \mid i \in I\rangle$ be subdirect, where $\{\theta_i^{\mathcal{B}} \mid i \in I\}$ is a set of irreducible congruences containing all completely meet irreducible congruences. Then $\mathcal{L}$ strongly patches over the dual ring of hull-kernel closed sets if and only if $\mathcal{L}$ is congruence permutable.

Proof: For every hull-kernel closed F, G, $\theta^{\mathcal{L}}_{F\cap G} = \theta^{\mathcal{L}}_{F} \vee \theta^{\mathcal{L}}_{G}$ by Lemma 3.16. The assertion follows from Lemma 4.35. ∎

An algebra is called arithmetical if it is congruence distributive and permutable (Pixley [33]).

Corollary 4.41 Let $\mathcal{L}$ be an arithmetical algebra and let $\mathcal{L} \subseteq \Pi\langle \mathfrak{A}_i \mid i \in I\rangle$ be subdirect, where $\{\theta^{\mathcal{L}}_i \mid i \in I\}$ is a set of irreducible congruences containing all completely meet irreducible congruences. Then $\mathcal{L}$ is normal over the ring of hull-kernel open sets.

Proof: Use Lemmas 3.1 and 4.40. ∎

Theorem 4.42 Let $\mathcal{L}$ be an arithmetical algebra and let $\mathcal{L} \subseteq \Pi\langle \mathfrak{A}_i \mid i \in I\rangle$ be subdirect, where $\{\theta^{\mathcal{L}}_i \mid i \in I\}$ is the set of all completely meet irreducible congruences on $\mathcal{L}$. Let $\mathcal{R}$ be the ring of hull-kernel open sets and let J be the set of prime filters on $\mathcal{R}$. Then $\mathcal{L} \cong g(\mathcal{L}) \subseteq \Pi\langle \mathcal{L}/\mathcal{U} \mid \mathcal{U} \in J\rangle$ is $\mathcal{T}(\mathcal{R}^*)$-global, where

(i) for every $i \in I$, $\mathcal{L}/h(i) \cong \mathfrak{A}_i$;

(ii) for every $\mathcal{U} \in J$, $\mathcal{L}/\mathcal{U} \in HSP_u\{\mathfrak{A}_i \mid i \in I\}$.

Proof: For the first part of the assertion use Theorem 4.13 and Corollary 4.41, and for the second use Corollary 4.15 and Theorem 4.23. ∎

Wolf [41] obtains a global subdirect representation for arithmetical algebras. His result is a typical sheaf representation theorem: Absolutely nothing is known about the factors. Our result is somewhat better because we do know something about the factors, and we have applications where this information turns out to be significant. We shall discuss these applications in Section 6.

We continue to consider congruence distributive algebras $\mathcal{L}$. A congruence ψ on $\mathcal{L}$ is called a factor congruence if there exists a congruence χ on $\mathcal{L}$ such that $\psi \cap \chi = \mathbb{O}^{\mathcal{L}}$ and $\psi\chi = \mathbb{1}^{\mathcal{L}}$ (that is $\mathcal{L} \cong \mathcal{L}/\psi \times \mathcal{L}/\chi$ under the canonical mapping). Since $\mathcal{L}$ is congruence distributive, the set of factor congruences of $\mathcal{L}$ determines a sublattice of the congruence lattice of $\mathcal{L}$ containing $\mathbb{O}^{\mathcal{L}}$ and $\mathbb{1}^{\mathcal{L}}$, which is a Boolean algebra of permuting congruences (Pierce [32], p. 88, Problem 11), so that Theorem 4.34 applies. We shall again use a subdirect representation $\mathfrak{L} \subseteq \Pi\langle \mathfrak{A}_i \mid i \in I\rangle$, where $\{\theta_i^{\mathfrak{L}} \mid i \in I\}$ is a set of irreducible congruences containing all completely meet irreducible congruences. Now let $\mathcal{R}$ be the ring of hull-kernel <u>closed</u> sets and let $\mathfrak{F}$ be the field of hull-kernel clopen sets.

<u>Lemma</u> 4.43 The anti isomorphism $F \in \mathcal{R} \to \theta_F^{\mathfrak{L}}$ maps $\mathfrak{F}$ onto the Boolean algebra of factor congruences of $\mathfrak{L}$. ∎

<u>Corollary</u> 4.44 $\mathfrak{L}$ patches over $\mathfrak{F}$.

<u>Proof</u>: Use Lemmas 4.30 and 4.43. ∎

<u>Theorem</u> 4.45 Let $\mathfrak{L}$ be a congruence distributive algebra and let $\mathfrak{L} \subseteq \Pi\langle \mathfrak{A}_i \mid i \in I\rangle$ be subdirect, where $\{\theta_i^{\mathfrak{L}} \mid i \in I\}$ is a set of irreducible congruences containing all completely meet irreducible congruences. Let $\mathfrak{F}$ be the field of hull-kernel clopen sets and let J be the set of prime filters on $\mathfrak{F}$. Then $\mathfrak{L} \cong g(\mathfrak{L}) \subseteq \Pi\langle \mathfrak{L}/\mathcal{U} \mid \mathcal{U} \in J\rangle$ is $\mathfrak{T}(\mathfrak{F}^*)$-global.

<u>Proof</u>: Use Theorem 4.13 and Corollary 4.44. ∎

Theorem 4.45 is rather deceptive because there may be very "few" hull-kernel clopen sets. Nevertheless, in this case we can say something about the factors of the global subdirect representation. First of all notice, since $\mathfrak{F}$ is a field of sets, for any $\mathcal{U} \in J$ and any $x, y \in B$,

$$x \equiv y(\mathcal{U}) \text{ iff for some } F \in \mathcal{U},\ F \subseteq E(x, y).$$

Thus

$$\equiv (\mathcal{U}) = \bigcup\{\theta_F^{\mathfrak{L}} \mid F \in \mathcal{U}\} = \theta_{\bigcap\mathcal{U}}^{\mathfrak{L}}$$

since the mapping $F \in \mathcal{R} \to \theta_F^{\mathfrak{L}}$ is an anti isomorphism between the ring $\mathcal{R}$ and the congruence lattice of $\mathfrak{L}$.

<u>Lemma</u> 4.46 Let $\mathcal{U} \in J$

(i) If $\mathcal{U} \notin h(I)$ then $\mathfrak{L}/\mathcal{U}$ is trivial.

(ii) $\mathfrak{L}/h(i) \cong \pi_{\bigcap h(i)}(\mathfrak{L})$.

<u>Proof</u>: (i) If $\mathcal{U} \notin h(I)$ then $\bigcap\mathcal{U} = \emptyset$. Now (i) follows from the remark above, and so does (ii). ∎

Remarks 4.47 In Theorems 4.42 and 4.45 we obtain entirely opposite situations. In Theorem 4.42 we obtain all old factors back but may lose track of the new factors. In Theorem 4.45 we obtain no non-trivial new factors but have to "blow up" the old factors beyond recognition. Moreover, in this case the process of "blowing up" the factors becomes particulary visible because the original subdirect representation is fully expanded. However, there is a special case where the "blown up" factors $\pi_{\bigcap h(i)}(\mathcal{B})$ are still "small" in some algebraically tangible fashion. Notice that the global subdirect representation of $\mathcal{B}$ in Theorem 4.45 is obtained from all direct decompositions of $\mathcal{B}$. An algebra $\mathfrak{A}$ is called directly indecomposable if $\mathbb{O}^{\mathfrak{A}}$ and $\mathbb{1}^{\mathfrak{A}}$ are its only factor congruences (that is, if $\mathfrak{A} \cong \mathfrak{A}_0 \times \mathfrak{A}_1$ then $\mathfrak{A} \cong \mathfrak{A}_0$ or $\mathfrak{A} \cong \mathfrak{A}_1$). We shall now determine when the factors of the global subdirect representation of $\mathcal{B}$ in Theorem 4.45 are directly indecomposable.

A topological space has the component separation property if any two connected components are separated by a clopen set. A compact Hausdorff space has the component separation property whereas a compact T_1-space may fail to have it. We shall show that the factors $\mathcal{B}/\mathcal{U}$ in Theorem 4.45 are directly indecomposable if and only if I has the component separation property.

Lemma 4.48 Suppose an algebra $\mathcal{C} \subseteq \Pi\langle \mathcal{C}_i \mid i \in I\rangle$ is a fully expanded subdirect product and let $F \subseteq I$, where $F = \overline{F}$. Then $\pi_F(\mathcal{C}) \subseteq \Pi\langle \mathcal{C}_i \mid i \in F\rangle$ is also fully expanded.

Proof: Let $\mathcal{D} = \pi_F(\mathcal{C})$ and let ψ be a completely meet irreducible congruence on $\mathcal{D}$. For $x, y \in C$ define

$$x\chi y \quad \text{iff} \quad \pi_F(x)\psi\pi_F(y)$$

Since $\mathcal{C}/\chi \cong \mathcal{D}/\psi$ are subdirectly irreducible and $\mathcal{C}$ is fully expanded, by Lemma 3.20, there exists $j \in I$ such that $\chi = \theta_j^{\mathcal{C}}$. We claim that $j \in F$ and $\psi = \theta_j^{\mathcal{D}}$. Indeed, suppose $j \notin F$. Since $F = \overline{F}$, there exist $x, y \in C$ such that $F \subseteq E(x, y)$ but $j \notin E(x, y)$. Thus $\pi_F(x) = \pi_F(y)$ so that $x\chi y$. But then $x(j) = y(j)$ which cannot be. This establishes that $j \in F$, and it follows that for every $x, y \in C$

$$\pi_F(x)\psi\pi_F(y) \quad \text{iff} \quad x(j) = y(j) \quad \text{iff} \quad \pi_F(x)\theta_j^{\mathcal{D}}\pi_F(y).$$

The assertion follows from Lemma 3.20. ∎

We return to the setting of Theorem 4.45 and Lemma 4.46.

Lemma 4.49 $\mathcal{B}/h(i)$ is directly indecomposable if and only if $\bigcap h(i)$

is connected.

Proof: By Lemma 4.46,

$$\mathcal{L}/h(i) \cong \pi_{\bigcap h(i)}(\mathcal{L}) \subseteq \Pi\langle \mathfrak{A}_j \mid j \in \bigcap h(i)\rangle$$

and by Lemma 4.48, this subdirect product is fully expanded. By Lemma 4.43, the field of clopen subsets of $\bigcap h(i)$ is anti isomorphic to the Boolean algebra of factor congruences of $\mathcal{L}/h(i)$. ∎

Lemma 4.50 For every $i \in I$, $\mathcal{L}/h(i)$ is directly indecomposable if and only if I has the component separation property.

Proof: For each $i \in I$, let C_i be the connected component of I containing i. Then $C_i \subseteq \bigcap h(i)$, and $C_i = \bigcap h(i)$ if and only if $\bigcap h(i)$ is connected. Thus by Lemma 4.49, $\bigcap h(i)$ is the connected component of i if and only if $\mathcal{L}/h(i)$ is directly indecomposable. On the other hand, I has the component separation property if and only if for every $i \in I$, $\bigcap h(i)$ is the connected component of i. ∎

Theorem 4.51 Let $\mathcal{L}$ be a congruence distributive algebra and let $\mathcal{L} \subseteq \Pi\langle \mathfrak{A}_i \mid i \in I\rangle$ be subdirect, where $\{\theta_i^{\mathcal{L}} \mid i \in I\}$ is a set of irreducible congruences containing all completely meet irreducible congruences. If I has the component separation property then $\mathcal{L}$ is isomorphic to a global subdirect product of directly indecomposable algeblas over a Boolean space.

Proof: Use Theorem 4.45 and Lemma 4.50. ∎

The component separation property is a severly restrictive requirement for the hull-kernel topology. Nevertheless, in Section 8 we shall exhibit large classes of lattice ordered rings and abelian groups to which Theorem 4.51 applies. It is interesting to notice, however, that in many applications the hull-kernel topology does not appear to enjoy strong separation properties. At the end of Section 3 we saw that we can always "make the hull-kernel topology T_o" by projecting onto a complete transversal. We conclude this section by showing that this process does not disturb patching over the ring of hull-kernel closed sets. In fact we shall show that, more generally, we can always make $\mathfrak{T}(\mathcal{R})$ a T_o-topology without disturbing patching over the dual ring $\breve{\mathcal{R}}$. There are again some subtle points involved that we should like to call to the reader's attention. Suppose $\mathcal{L} \subseteq \Pi\langle \mathfrak{A}_i \mid i \in I\rangle$ is an algebra, $\mathcal{R}$ is a ring of subsets of I and $J \subseteq I$ is a complete $\mathcal{L}$-transversal. If $\rho: I \to J$

is the retraction defined at the end of Section 2 and we define

$$\mathcal{R}/\rho = \{G \subseteq J \mid \rho^{-1}(G) \in \mathcal{R}\},$$

then $\mathcal{R}/\rho$ is a ring of subsets of J and we may endow J with three topologies: the relative topology $\mathcal{T}(\mathcal{R} \mid J)$, the quotient topology $\mathcal{T}(\mathcal{R})/\rho$ and the topology $\mathcal{T}(\mathcal{R}/\rho)$ with basis $\mathcal{R}/\rho$ of closed sets. Now we obtain

$$\mathcal{T}(\mathcal{R}/\rho) \subseteq \mathcal{T}(\mathcal{R})/\rho \subseteq \mathcal{T}(\mathcal{R} \mid J)$$

and these topologies may all be distinct!

Lemma 4.52 Suppose $\mathcal{B} \subseteq \Pi\langle \mathfrak{A}_i \mid i \in I\rangle$ is an algebra, $\mathcal{R}$ is a ring of subsets of I and $J \subseteq I$ is a complete $\mathcal{B}$-transversal. If $\mathcal{B}$ patches over $\breve{\mathcal{R}}$ (is normal over $\mathcal{R}$) then $\pi_J(\mathcal{B})$ patches over $\breve{\mathcal{R}/\rho}$ (is normal over $\mathcal{R}/\rho$).

Proof: Suppose $\mathcal{B}$ patches over $\breve{\mathcal{R}}$ and let $r < \omega$, $F_p \in \mathcal{R}/\rho$ and $x_p \in B$, where $\bigcap_{r<p} F_p = \emptyset$ and $D(\pi_J(x_p), \pi_J(x_q)) \subseteq F_p \cup F_q$ whenever $p, q < r$. Then $\bigcap_{p<r} \rho^{-1}(F_p) = \emptyset$ and by Lemma 2.37, $D(x_p, x_q) \subseteq \rho^{-1}(F_p) \cup \rho^{-1}(F_q)$ whenever $p, q < r$. By hypothesis, there exists $y \in B$ such that $D(y, x_p) \subseteq \rho^{-1}(F_p)$ for all $p < r$, and therefore $D(\pi_J(y), \pi_J(x_p)) \subseteq F_p$ for all $p < r$. We have shown that $\pi_J(\mathcal{B})$ patches over $\breve{\mathcal{R}/\rho}$, and the remainder of the assertion follows from Lemma 2.37. ∎

Corollary 4.53 Suppose an algebra $\mathcal{B} \subseteq \Pi\langle \mathfrak{A}_i \mid i \in I\rangle$ is subdirect and induces a hull-kernel topology on I. Let $\mathcal{R}$ be the ring of hull-kernel closed sets and let $J \subseteq I$ be a complete $\mathcal{B}$-transversal. If $\mathcal{B}$ patches over $\breve{\mathcal{R}}$ (is (semi) normal over $\mathcal{R}$) then $\pi_J(\mathcal{B})$ patches over $\breve{\mathcal{R} \mid J}$ (is (semi) normal over $\mathcal{R} \mid J$).

Proof: If $\mathcal{R}$ is the ring of hull-kernel closed sets then $\mathcal{R}/\rho = \mathcal{R} \mid J$. ∎

It is important to realize that projections in general do not preserve patching over a ring. There is one notable exception which is easy to prove.

Lemma 4.54 Suppose $\mathcal{B} \subseteq \Pi\langle \mathfrak{A}_i \mid i \in I\rangle$ and $\mathfrak{F}$ is a field of subsets of I. If $\mathcal{B}$ patches over $\mathfrak{F}$ (is (semi) normal over $\mathfrak{F}$) and $J \subseteq I$ then $\pi_J(\mathcal{B})$ patches over $\mathfrak{F} \mid J$ (is (semi) normal over $\mathfrak{F} \mid J$.) ∎

5. CLOSURE UNDER THE NORMAL TRANSFORM

In this section we shall investigate the notion of "tightness" which has the most interesting applications: normalcy over a field. It turns out that this notion has a very simple internal characterization in terms of the normal transform.

Lemma 5.1 Let $\mathfrak{B} \subseteq \Pi\langle \mathfrak{A}_i \mid i \in I\rangle$ and let $\mathfrak{F}$ be a field of subsets of I. Then $\mathfrak{B}$ is normal over $\mathfrak{F}$ if and only if

(i) for every $x, y \in B$ and every $F \in \mathfrak{F}$, if

$$z(i) = \begin{cases} x(i) & \text{for } i \in F \\ y(i) & \text{for } i \notin F \end{cases}$$

then $z \in B$;

(ii) for every atomic formula φ and every $x \in B^n$, $[\![\varphi(x)]\!] \in \mathfrak{F}$.

Proof: Use Lemma 4.5. ∎

A (locally) Boolean space is a zero-dimensional (locally) compact Hausdorff space. $\mathfrak{B}$ is called a (locally) Boolean substructure of $\Pi\langle \mathfrak{A}_i \mid i \in I\rangle$ if $\mathfrak{B}$ is a global substructure of $\Pi\langle \mathfrak{A}_i \mid i \in I\rangle$, where I is a (locally) Boolean space and for every atomic formula φ and every $x \in B^n$, $[\![\varphi(x)]\!]$ is clopen. If, moreover, $\mathfrak{B}$ is subdirect then $\mathfrak{B}$ is called a (locally) Boolean subdirect product of $\{\mathfrak{A}_i \mid i \in I\}$, and if $\mathfrak{A}_i = \mathfrak{A}$ for all $i \in I$ then $\mathfrak{B}$ is called a (locally) Boolean subdirect power of $\mathfrak{A}$.

Lemma 5.2 If $\mathfrak{B} \subseteq \Pi\langle \mathfrak{A}_i \mid i \in I\rangle$ is Boolean then $\mathfrak{B}$ is normal over the field of clopen sets.

Proof: Use Lemma 4.4. ∎

Lemma 5.3 Suppose $\mathfrak{B} \subseteq \Pi\langle \mathfrak{A}_i \mid i \in I\rangle$ and $\mathfrak{F}$ is a field of subsets of I. Let $\mathfrak{T}(\mathfrak{F})$ be the topology on I with basis of clopen sets $\mathfrak{F}$. If $\mathfrak{B}$ is normal over $\mathfrak{F}$ and I is a compact Hausdorff space then $\mathfrak{B}$ is Boolean.

Proof: Use Lemma 4.4. ∎

Lemma 5.4 Let $\mathcal{B} \subseteq \Pi\langle \mathfrak{A}_i \mid i \in I\rangle$ and let $\mathcal{F}$ be a field of subsets of I. If $\mathcal{B}$ is normal over $\mathcal{F}$ then $\mathcal{B}$ is closed under $\bar{n}$.

Proof: Let $x, y, z, w \in B$ and define

$$u(i) = \begin{cases} z(i) & \text{if } i \in E(x, y) \\ w(i) & \text{if } i \notin E(x, y) \end{cases}$$

Then $u = \bar{n}(x, y, z, w)$. ∎

The next observation is due to Keimel and Werner [25].

Lemma 5.5 For all $x, y, u, v \in \Pi\langle A_i \mid i \in I\rangle$,

(i) $E(x, y) \cap E(u, v) = E(\bar{n}(x, y, u, x), \bar{n}(y, x, v, y))$

(ii) $E(x, y) \cap D(u, v) = D(\bar{n}(x, y, u, x), \bar{n}(x, y, v, x))$

(iii) $D(x, y) \cap D(u, v) = D(\bar{n}(x, y, x, u), \bar{n}(x, y, x, v))$. ∎

Corollary 5.6 Suppose an algebra $\mathcal{B} \subseteq \Pi\langle \mathfrak{A}_i \mid i \in I\rangle$ is closed under $\bar{n}$. Then

(i) $$\mathcal{R}_o = \{D(x, y) \mid x, y \in B\}$$
is a Boolean ring of subsets of I;

(ii) $$\breve{\mathcal{R}}_o = \{E(x, y) \mid x, y \in B\}$$
is a basis for the equalizer topology on I induced by $\mathcal{B}$;

(iii) $$\mathcal{F}_o = \mathcal{R}_o \cup \breve{\mathcal{R}}_o$$
is a field of subsets of I;

(iv) $\mathcal{B}$ is normal over $\mathcal{F}_o$;

(v) $\mathcal{B}$ is normal over $\mathcal{R}_o$.

Proof: (i) - (iii) follow from Lemma 5.5 and (iv) follows from Lemma 5.1. Finally, (v) follows from (iv). ∎

Hence a subdirect product is normal over some field if and only if it is closed under $\bar{n}$.

Lemma 5.7 Suppose an algebra $\mathcal{B} \subseteq \Pi\langle \mathfrak{A}_i \mid i \in I\rangle$ and $\mathcal{T}$ is a topology on I. If $\mathcal{B}$ is normal over the ring of all closed sets then the following are equivalent:

(i) For every $x, y \in B$, $E(x, y)$ is clopen.

(ii) $\mathcal{B}$ is normal over the field of clopen sets.

Moreover, (ii) implies

(iii) $\mathcal{B}$ is closed under $\bar{n}$,

and if there exist $x, y \in B$ such that $I = D(x, y)$ then (iii) implies (ii).

Proof: The equivalence of (i) and (ii) is obvious and (ii) implies (iii) by Lemma 5.4. Finally, assume (iii) and let $x, y \in B$, where $I = D(x, y)$. By Corollary 5.6, for every $u, v \in B$,

$$E(u, v) = D(x, y) \cap E(u, v) = D(x, y) - D(u, v)$$

is closed. ∎

Corollary 5.8 If an algebra $\mathcal{B} \subseteq \Pi\langle \mathfrak{A}_i \mid i \in I\rangle$ is $\mathcal{T}$-global then the following are equivalent:

(i) For every $x, y \in B$, $E(x, y)$ is clopen.

(ii) $\mathcal{B}$ is normal over the field of clopen sets.

Moreover, (ii) implies

(iii) $\mathcal{B}$ is closed under $\bar{n}$,

and if there exist $x, y \in B$ such that $I = D(x, y)$ then (iii) implies (ii).

Proof: Use Lemmas 4.3 and 5.7. ∎

Now let $\mathcal{B} \subseteq \Pi\langle \mathfrak{A}_i \mid i \in I\rangle$ be subdirect and let $\mathfrak{F}$ be a field of subsets of I. Assume that $\mathcal{B}$ is normal over $\mathfrak{F}$ and let $\mathcal{T}(\mathfrak{F})$ be the topology on I with basis of clopen sets $\mathfrak{F}$. Again in general $\mathcal{B}$ will not be $\mathcal{T}(\mathfrak{F})$-global, however the process of globalizing $\mathcal{B}$ now yields much more information because every prime filter on $\mathfrak{F}$ is maximal. Again let J be the set of prime filters on $\mathfrak{F}$. Recall that we now obtain for $\mathcal{U} \in J$ and $x, y \in B$

$$x \equiv y(\mathcal{U}) \quad \text{iff} \quad E(x, y) \in \mathcal{U}$$

Globalizing $\mathcal{B}$ we construct the Gelfand mapping $g: \mathcal{B} \to \Pi\langle \mathcal{B}/\mathcal{U} \mid \mathcal{U} \in J\rangle$ as in Section 4.

Theorem 5.9 Suppose $\mathcal{B} \subseteq \Pi\langle \mathfrak{A}_i \mid i \in I\rangle$ is subdirect and $\mathfrak{F}$ is a field of subsets of I. If $\mathcal{B}$ is normal over $\mathfrak{F}$ then $\mathcal{B} \cong g(\mathcal{B}) \subseteq \Pi\langle \mathcal{B}/\mathcal{U} \mid \mathcal{U} \in J\rangle$ is Boolean. Moreover,

(i) for every $i \in I$, $\mathcal{B}/h(i) \cong \mathfrak{A}_i$;

(ii) for every $\mathcal{U} \in J$, $\mathcal{B}/\mathcal{U} \in O\{\mathfrak{A}_i \mid i \in I\}$;

(iii) if $D(x, y)$ is compact for every $x, y \in B$, then for every $\mathcal{U} \in J$, $\mathcal{B}/\mathcal{U}$ is trivial or there exists $i \in I$ such that $\mathcal{U} = h(i)$.

Proof: Use Corollary 4.21 and Theorem 4.25. ∎

Corollary 5.10 Every subdirect product which is normal over a field $\mathfrak{F}$ of subsets of the index set is isomorphic to a Boolean subdirect product over a space with a clopen basis isomorphic to $\mathfrak{F}$. ∎

Corollary 5.11 Suppose an algebra $\mathfrak{B} \subseteq \Pi\langle \mathfrak{A}_i \mid i \in I\rangle$ is subdirect. If $\mathfrak{B}$ is closed under $\bar{n}$ then $\mathfrak{B}$ is isomorphic to a Boolean subdirect product $\mathfrak{C} \subseteq \Pi\langle \mathfrak{B}_j \mid j \in J\rangle$, where

(i) for every $i \in I$ there exists $j \in J$ such that $\mathfrak{A}_i \cong \mathfrak{B}_j$.

(ii) for every $j \in J$, $\mathfrak{B}_j \in O\{\mathfrak{A}_i \mid i \in I\}$.

Proof: Use Corollary 5.6 and Theorem 5.9. ∎

A more careful investigation of closure under the normal transform will yield sharper results than Corollary 5.11.

Lemma 5.12 Suppose an algebra $\mathfrak{B} \subseteq \Pi\langle \mathfrak{A}_i \mid i \in I\rangle$ is subdirect. If every $\mathfrak{A}_i$ is non-trivial and $\mathfrak{B}$ is closed under $\bar{n}$ then $\mathfrak{B}$ induces a hull-kernel topology on I.

Proof: We shall show that for every $F, G \subseteq I$, $\overline{F \cup G} \subseteq \overline{F} \cup \overline{G}$. Let $i \in \overline{F \cup G}$, where $i \notin \overline{F}$. Consider $u, v \in B$ such that $G \subseteq E(u, v)$. Since $i \notin \overline{F}$ there are $x, y \in B$ such that $F \subseteq E(x, y)$ but $x(i) \neq y(i)$. Let $z = \bar{n}(x, y, u, v)$. Then $F \subseteq E(z, u)$ since $F \subseteq E(x, y)$, and $G \subseteq E(z, u)$ since $G \subseteq E(u, v)$. Thus $F \cup G \subseteq E(z, u)$ and therefore

$$i \in \overline{F \cup G} \subseteq E(z, u).$$

Since $x(i) \neq y(i)$, $v(i) = z(i) = u(i)$ and therefore $i \in \overline{G}$. The assertion follows from Lemma 3.4. ∎

Lemma 5.13 Suppose an algebra $\mathfrak{B} \subseteq \Pi\langle \mathfrak{A}_i \mid i \in I\rangle$ is subdirect. If every $\mathfrak{A}_i$ is non-trivial and $\mathfrak{B}$ is closed under $\bar{n}$ then for every $x, y \in B$, $D(x, y)$ is hull-kernel clopen.

Proof: Let $i \in E(x, y)$. Since $\mathfrak{A}_i$ is non-trivial, there exist $u, v \in B$ such that $i \in D(u, v)$. Thus

$$i \in D(u, v) \cap E(x, y) = D(u, v) - D(x, y) \subseteq E(x, y).$$

By Corollaries 5.6 and 3.5, $E(x, y)$ is hull-kernel open. The assertion follows from Corollary 3.5. ∎

Corollary 5.14 Suppose an algebra $\mathfrak{B} \subseteq \Pi\langle \mathfrak{A}_i \mid i \in I\rangle$ is subdirect. If every $\mathfrak{A}_i$ is non-trivial and $\mathfrak{B}$ is closed under $\bar{n}$ then

(i) $\mathfrak{T}(\mathfrak{R}_o)$ (that is the topology with basis of closed sets $\mathfrak{R}_o$) is the equalizer topology induced by $\mathfrak{B}$;

(ii) $\mathfrak{T}(\mathfrak{F}_o)$ (that is the topology with basis of clopen sets $\mathfrak{F}_o$) is the hull-kernel topology induced by $\mathfrak{B}$.

Proof: (i) follows from Corollary 5.6. (ii) By Lemma 5.13, every $F \in \mathfrak{F}_o$ is hull-kernel clopen. Thus every $\mathfrak{T}(\mathfrak{F}_o)$-closed set is hull-kernel closed. Conversely, every $F \in \breve{\mathfrak{R}}_o$ is $\mathfrak{T}(\mathfrak{F}_o)$-clopen. By Corollary 3.5, every hull-kernel closed set is $\mathfrak{T}(\mathfrak{F}_o)$-closed. ∎

Lemma 5.15 Suppose an algebra $\mathfrak{B} \subseteq \Pi\langle \mathfrak{A}_i \mid i \in I\rangle$ is fully expanded and subdirect. If every $\mathfrak{A}_i$ is non-trivial and $\mathfrak{B}$ is closed under $\bar{n}$ then

(i) the equalizer topology induced by $\mathfrak{B}$ is compact;

(ii) the hull-kernel topology induced by $\mathfrak{B}$ is zero dimensional and locally compact.

Proof: (ii) follows from Corollaries 3.26 and 5.14. To prove (i), suppose $x_\xi, y_\xi \in B$ for each $\xi \in \Delta$ where

$$I = \bigcup_{\xi\in\Delta} E(x_\xi, y_\xi).$$

Choose $\delta \in \Delta$. Then

$$I = E(x_\delta, y_\delta) \cup \bigcup\{E(x_\xi, y_\xi) \mid \xi \in \Delta, \xi \neq \delta\}$$

and

$$D(x_\delta, y_\delta) \subseteq \bigcup\{E(x_\xi, y_\xi) \mid \xi \in \Delta, \xi \neq \delta\}.$$

By Lemma 3.25, $D(x_\delta, y_\delta)$ is compact in the hull-kernel topology. The assertion follows from Corollary 5.14. ∎

Lemma 5.16 Suppose an algebra $\mathfrak{B} \subseteq \Pi\langle \mathfrak{A}_i \mid i \in I\rangle$ is fully expanded and subdirect. If every $\mathfrak{A}_i$ is non-trivial and $\mathfrak{B}$ is closed under $\bar{n}$ then the following are equivalent:

(i) The equalizer topology induced by $\mathfrak{B}$ is the same as the hull-kernel topology induced by $\mathfrak{B}$.

(ii) The hull-kernel topology induced by $\mathfrak{B}$ is compact.

(iii) For some $x, y \in B$, $D(x, y) = I$.

Proof: (i) implies (ii) by Lemma 5.15 and (ii) implies (iii) by Corollaries 3.26 and 5.6. Finally, (iii) implies (i) by Corollary 5.14. ∎

Theorem 5.17 Suppose an algebra $\mathcal{B} \subseteq \Pi\langle \mathfrak{A}_i \mid i \in I\rangle$ is fully expanded and subdirect. If every $\mathfrak{A}_i$ is non-trivial and $\mathcal{B}$ is closed under $\bar{n}$ then

(i) $\mathcal{B}$ is global;

(ii) if for some $x, y \in B$, $D(x, y) = I$ then $\mathcal{B}$ is global with respect to the hull-kernel topology induced by $\mathcal{B}$; moreover, $\mathcal{B}$ is isomorphic to a Boolean subdirect product of $\{\mathfrak{A}_j \mid j \in J\}$, where J is a complete $\mathcal{B}$-transversal;

(iii) if $D(x, y) \neq I$ for all $x, y \in B$ then $\mathcal{B}$ is isomorphic to a Boolean subdirect product of $\{\mathfrak{A}_j \mid j \in J \cup \{j_o\}\}$, where $J \subseteq I$ and $\mathfrak{A}_{j_o}$ is trivial.

Proof: To prove (i) and the first part of (ii), use Lemma 4.4, Corollaries 5.6 and 5.14, and Lemmas 5.15 and 5.16. The second part of (ii) follows from Theorem 2.39. To prove (iii), recall that $\mathcal{B}$ is normal over $\mathfrak{F}_o$ and globalize $\mathcal{B}$. By Lemma 1.12 and Corollary 4.15, there is exactly one new factor $\mathcal{B}/\breve{\mathcal{R}}$, which is trivial. ∎

Case (iii) of Theorem 5.17 is usually treated differently in the literature. To see what is involved, suppose an algebra $\mathcal{B} \subseteq \Pi\langle \mathfrak{A}_i \mid i \in I\rangle$ is subdirect and fully expanded, where every $\mathfrak{A}_i$ is non-trivial and $\mathcal{B}$ is closed under $\bar{n}$. If $D(x, y) \neq I$ for all $x, y \in B$ then a global subdirect representation for $\mathcal{B}$ is obtained by globalizing $\mathcal{B}$. In this case the compactification of the index space requires the adjunction of a trivial factor. However, it is also possible to characterize $\mathcal{B}$ as a subalgebra of its global closure, and this is usually done in the literature. It is interesting to notice that the global closure of $\mathcal{B}$ is not in general isomorphic to its globalization, and we shall illustrate this with biregular rings in Section 7.

Theorem 5.18 Suppose an algebra $\mathcal{B} \subseteq \Pi\langle \mathfrak{A}_i \mid i \in I\rangle$ is fully expanded and subdirect. Suppose every $\mathfrak{A}_i$ is non-trivial and $\mathcal{B}$ is closed under $\bar{n}$. Let $\mathfrak{T}$ be the hull-kernel topology induced by $\mathcal{B}$ and let $b \in B$ be fixed. Then

(i) I is a zero-dimensional, locally compact space;

(ii) I is Hausdorff if and only if $\mathcal{B}$ is irredundant;

(iii) $x \in B$ if and only if $x \in \Gamma(B, \mathfrak{T})$ and $D(x, b)$ is compact.

Proof: (i) follows from Lemma 5.15, and (ii) follows from (i) and Lemma

3.30. To prove (iii), suppose $x \in B$. Then by Lemma 3.25, $D(x, b)$ is compact. Conversely, suppose $x \in \Gamma(B, \mathfrak{T})$ and $D(x, b)$ is compact. For every $i \in D(x, b)$ there exists $y_i \in B$ such that $i \in E(x, y_i)$. Thus

$$D(x, b) \subseteq \bigcup\{E(x, y_i) \mid i \in D(x, b)\},$$

where $E(x, y_i)$ is open for all $i \in I$. Hence there exist $r < \omega$ and $u_p, v_p \in B$ and $i_p \in D(x, b)$, for each $p < r$, such that

$$D(x, b) \subseteq \bigcup_{p<r} D(u_p, v_p)$$

and $D(u_p, v_p) \subseteq E(x, y_{i_p})$ for all $p < r$. By Corollary 5.6 there exist $u_r, v_r \in B$ such that

$$D(u_r, v_r) = \bigcup_{p<r} D(u_p, v_p).$$

Thus

$$I = E(u_r, v_r) \cup \bigcup_{p<r} D(u_p, v_p)$$

and

$$D(u_p, v_p) \cap D(u_q, v_q) \subseteq E(x, y_{i_p}) \cap E(x, y_{i_q}) \subseteq E(y_{i_p}, y_{i_q})$$

whenever $p, q < r$. Hence there exists $y \in B$ such that $E(u_r, v_r) \subseteq E(y, x)$ and for each $p < r$,

$$D(u_p, v_p) \subseteq E(y, y_{i_p}) \cap E(x, y_{i_p}) \subseteq E(y, x).$$

It follows that $x = y \in B$. This establishes (iii). ∎

In applications of Theorem 5.18 in the literature I is usually a locally Boolean space. In Section 3 we saw that we can always "make the hull-kernel topology T_0" without disturbing full expansion. It is easy to see that this process does not disturb closure under the normal transform either.

<u>Lemma</u> 5.19 If $\mathcal{L} \subseteq \Pi\langle \mathfrak{A}_i \mid i \in I\rangle$ is closed under $\bar{n}$ and $J \subseteq I$ then $\pi_J(\mathcal{L}) \subseteq \Pi\langle \mathfrak{A}_j \mid j \in J\rangle$ is closed under $\bar{n}$. ∎

Lemma 5.19 tells us that in Theorem 5.18 we may assume that I actually is a locally Boolean space. Notice, however, that in Theorem 5.18 we do not in general obtain $E(x, y)$ clopen for all $x, y \in \Gamma(B, \mathfrak{T})$. In other words, $\Gamma(\mathcal{L}, \mathfrak{T})$ is <u>not</u> in general a locally Boolean subdirect product.

Next we shall again consider congruence distributive algebras $\mathcal{L} \subseteq \Pi\langle \mathfrak{A}_i \mid i \in I\rangle$, where $\{\theta_i^{\mathcal{L}} \mid i \in I\}$ is a set of irreducible congruences containing all completely meet irreducible congruences (See Theorems 4.42 and 4.45). Notice that for any congruence ψ on $\mathcal{L}$, $\psi \cap \bar{\psi} = \mathbb{0}^{\mathcal{L}}$ and that

$\psi\bar{\psi} = \mathbb{1}^{\mathcal{B}}$ iff ψ is a factor congruence, where $\bar{\psi}$ is the pseudo complement of ψ

Lemma 5.20 Let $\mathcal{B}$ be a congruence distributive algebra and let $\mathcal{B} \subseteq \Pi\langle \mathfrak{A}_i \mid i \in I\rangle$ be subdirect, where $\{\theta_i^{\mathcal{B}} \mid i \in I\}$ is a set of irreducible congruences containing all completely meet irreducible congruences. Then for every $x, y \in B$,

$$\theta^{\mathcal{B}}(x, y) = \theta^{\mathcal{B}}_{E(x, y)} \quad \text{and} \quad \theta^{\mathcal{B}}(x, y)^{\sim} = \theta^{\mathcal{B}}_{D(x, y)}.$$

Proof: The first part of the assertion has been observed before (see the proof of Lemma 3.25), we verify the second.

$$\begin{aligned}\theta^{\mathcal{B}}(x, y)^{\sim} &= \bigcup\{\theta^{\mathcal{B}}_{\bar{J}} \mid \theta^{\mathcal{B}}_{\bar{J}} \cap \theta^{\mathcal{B}}_{E(x, y)} = \mathbb{0}^{\mathcal{B}}\} \\ &= \bigcup\{\theta^{\mathcal{B}}_{\bar{J}} \mid \bar{J} \cup E(x, y) = I\} \\ &= \theta^{\mathcal{B}}_{\overline{D(x, y)}} = \theta^{\mathcal{B}}_{D(x, y)}\end{aligned}$$

because

$$\overline{D(x, y)} = \bigcap\{\bar{J} \mid D(x, y) \subseteq \bar{J}\}. \blacksquare$$

Following Keimel [24], we call an algebra quasiregular if every principal congruence is a factor congruence.

Lemma 5.21 Let $\mathcal{B}$ be a congruence distributive algebra and let $\mathcal{B} \subseteq \Pi\langle \mathfrak{A}_i \mid i \in I\rangle$ be subdirect, where $\{\theta_i^{\mathcal{B}} \mid i \in I\}$ is a set of irreducible congruences containing all completely meet irreducible congruences. Then $\mathcal{B}$ is quasiregular if and only if $\mathcal{B}$ is closed under $\bar{n}$.

Proof: $\mathcal{B}$ is quasiregular if and only if for all $x, y \in B$, $\theta^{\mathcal{B}}(x, y)\,\theta^{\mathcal{B}}(x, y)^{\sim} = \mathbb{1}^{\mathcal{B}}$, if and only if for all $x, y \in B$, $\theta^{\mathcal{B}}_{E(x, y)}\,\theta^{\mathcal{B}}_{D(x, y)} = \mathbb{1}^{\mathcal{B}}$, if and only if $\mathcal{B}$ is closed under $\bar{n}$. $\blacksquare$

Next we generalize a fact which Keimel proves for lattice ordered rings:

Lemma 5.22 Let $\mathcal{B}$ be a congruence distributive quasiregular algebra. Then every subdirectly indecomposable homomorphic image of $\mathcal{B}$ is simple.

Proof: We must show that every meet irreducible congruence ψ on $\mathcal{B}$ is maximal. Choose $\langle x, y\rangle \notin \psi$ where $x, y \in B$. Now $\theta^{\mathcal{B}}(x, y) \cap \theta^{\mathcal{B}}(x, y)^{\sim} = \mathbb{0}^{\mathcal{B}} \subseteq \psi$ and since $\theta^{\mathcal{B}}(x, y) \nsubseteq \psi$ and ψ is meet irreducible, $\theta^{\mathcal{B}}(x, y)^{\sim} \subseteq \psi$. Since $\mathcal{B}$ is quasiregular,

$$\theta^{\mathcal{B}}(x, y) \vee \psi \supseteq \theta^{\mathcal{B}}(x, y) \vee \theta^{\mathcal{B}}(x, y)^{\sim} = \mathbb{1}^{\mathcal{B}}$$

so $\mathbb{1}^{\mathcal{B}}$ is the smallest congruence containing ψ and $\langle x, y\rangle$. ∎

Lemma 5.23 Let $\mathcal{B} \subseteq \Pi\langle \mathfrak{A}_i \mid i \in I\rangle$ be a congruence distributive algebra which is a global subdirect product of simple algebras over a compact space. Then $\mathcal{B}$ is fully expanded.

Proof: Suppose θ is a completely meet irreducible congruence on $\mathcal{B}$. By Jonsson [23], there is an ultrafilter $\mathcal{U}$ on I such that $x\theta y$ whenever $x, y \in B$ and $E(x, y) \in \mathcal{U}$. Since I is compact, $\mathcal{U}$ converges to some $j \in I$. Thus for any $x, y \in B$, if $x\theta_j^{\mathcal{B}}y$ then $j \in E(x, y)$ and therefore $E(x, y) \in \mathcal{U}$. Thus $x\theta y$ and we have shown that $\theta_j^{\mathcal{B}} \subseteq \theta$. Since $\theta_j^{\mathcal{B}}$ is maximal, $\theta_j^{\mathcal{B}} = \theta$ and the assertion follows from Lemma 3.20. ∎

Theorem 5.24 Let $\mathcal{B}$ be a congruence distributive algebra. Then $\mathcal{B}$ is quasi regular if and only if $\mathcal{B}$ is isomorphic to a Boolean subdirect product where each factor is simple or trivial, and at most one factor is trivial.

Proof: Suppose $\mathcal{B}$ is quasiregular and let $\mathcal{B} \subseteq \Pi\langle \mathfrak{A}_i \mid i \in I\rangle$ be subdirect, where $\{\theta_i^{\mathcal{B}} \mid i \in I\}$ is the set of all completely meet irreducible congruences. By Theorem 3.24 and Lemma 5.21, $\mathcal{B}$ is closed under $\bar{n}$ and the desired conclusion follows from Theorem 5.17 and Lemma 5.22.

Conversely, suppose $\mathcal{B} \subseteq \Pi\langle \mathfrak{A}_i \mid i \in I\rangle$ is a Boolean subdirect product, where each $\mathfrak{A}_i$ is simple or trivial, and at most one $\mathfrak{A}_i$ is trivial. By Lemmas 5.2 and 5.4, $\mathcal{B}$ is closed under $\bar{n}$. If every $\mathfrak{A}_i$ is non-trivial then by Lemma 5.23 $\{\theta_i^{\mathcal{B}} \mid i \in I\}$ is the set of all completely meet irreducible congruences and by Lemma 5.21, $\mathcal{B}$ is quasiregular. Otherwise, let $\mathfrak{A}_{i_o}$ be the unique trivial factor and let $J = I - \{i_o\}$. Then $\mathcal{B} \cong \pi_J(\mathcal{B}) \subseteq \Pi\langle \mathfrak{A}_j \mid j \in J\rangle$ is fully expanded and closed under $\bar{n}$ so that we can argue as in the first case. ∎

Closure under the normal transform also plays a role in several constructions discussed in Section 2. First we consider weak direct products.

Lemma 5.25 Suppose an algebra $\mathcal{B}$ is a weak direct product of $\{\mathfrak{A}_i \mid i \in I\}$. Then $\mathcal{B}$ is closed under $\bar{n}$. If, moreover, every $\mathfrak{A}_i$ is non-trivial then

(i) $\mathcal{R}_o$ is the (Boolean) ring of all finite subsets of I;

(ii) $\mathcal{F}_o = \mathcal{R}_o \cup \breve{\mathcal{R}}_o$ is the field of all finite or cofinite subsets of I;

(iii) the equalizer topology induced by $\mathcal{B}$ is the topology of cofinite subsets of I; moreover, with the equalizer topology I is a compact T_1 space;

(iv) the hull-kernel topology induced by $\mathcal{B}$ is the discrete topology.

Proof: If $x, y, z, w \in B$ then

$$E(x, y) \subseteq E(\bar{n}(x, y, z, w), z) \in \mathfrak{T}.$$

Thus $\mathcal{B}$ is closed under $\bar{n}$. The remainder of the assertion follows from Corollary 5.14 and Lemma 2.15. ∎

Theorem 5.26 Suppose an algebra $\mathcal{B}$ is a weak direct product of $\{\mathfrak{A}_i \mid i \in I\}$, where every $\mathfrak{A}_i$ is non-trivial. If I is infinite then $\mathcal{B}$ is isomorphic to a Boolean subdirect product of $\{\mathfrak{A}_i \mid i \in I \cup \{i_o\}\}$, where $\mathfrak{A}_{i_o}$ is trivial.

Proof: Recall that $\mathcal{B}$ is semi normal over $\mathfrak{F}_o$ and globalize $\mathcal{B}$. Then use Lemmas 1.12 and 5.25. ∎

Notice that Theorem 5.26 is an analogue of Theorem 5.17. We also obtain a somewhat trivial but illuminating analogue of Theorem 5.18.

Corollary 5.27 Suppose an algebra $\mathcal{B}$ is a weak direct product of $\{\mathfrak{A}_i \mid i \in I\}$, where every $\mathfrak{A}_i$ is non-trivial. Let $\mathfrak{T}$ be the hull-kernel topology induced by $\mathcal{B}$ and let $b \in B$ be fixed. Then

(i) I is a zero-dimensional locally compact Hausdorff space;

(ii) $\Gamma(B, \mathfrak{T}) = \Pi\langle \mathfrak{A}_i \mid i \in I\rangle$;

(iii) $x \in B$ if and only if $x \in \Gamma(B, \mathfrak{T})$ and $D(x, b)$ is compact.

Proof: By Lemma 5.25, the hull-kernel topology induced by $\mathcal{B}$ is the discrete topology. The assertion follows. ∎

Theorems 5.17 and 5.26 are rather interesting illustrations of the interplay between globality and normalcy. In both cases $\mathcal{B}$ is normal over the Boolean ring $\mathcal{R}_o$ and $\mathcal{R}_o$ is a basis of closed sets for the equalizer topology induced by $\mathcal{B}$. Since the equalizer topology is compact, $\mathcal{B}$ is a global subdirect product. But $\mathcal{B}$ is even normal over the still larger field $\mathfrak{F}_o = \mathcal{R}_o \cup \breve{\mathcal{R}}_o$, and $\mathfrak{F}_o$ is a (clopen) basis for the hull-kernel topology induced by $\mathcal{B}$. However, $\mathcal{B}$ is not in general global with respect to the hull-kernel topology, because the hull-kernel topology is not compact. We have to globalize $\mathcal{B}$. In this process we obtain all old factors back and adjoin at most one new factor, which is trivial. Moreover, we now have a Boolean subdirect product.

Next we shall see that after "removing redundancies" a Boolean subdirect product determines the field of clopen sets. These observations are essentially due to Werner [39].

Lemma 5.28 Suppose $\mathcal{B} \subseteq \Pi\langle \mathfrak{A}_i \mid i \in I\rangle$ is subdirect and $\mathfrak{F}$ is a field of subsets of I. Let $\mathfrak{T}(\mathfrak{F})$ be the topology on I with basis of clopen sets $\mathfrak{F}$ and let

$$E_o = \bigcap\{E(x, y) \mid x, y \in B\}$$

If $\mathcal{B}$ is normal over $\mathfrak{F}$ and I is compact then for any $F \in \mathfrak{F}$,

(i) if $F \cap E_o = \emptyset$ then there exist $x, y \in B$ such that $F = D(x, y)$;

(ii) if $E_o \subseteq F$ then there exist $x, y \in B$ such that $F = E(x, y)$.

Proof: (i) For every $i \in F$ there exist $x_i, y_i \in B$ such that $i \in D(x_i, y_i)$. Thus

$$F \subseteq \bigcup_{i \in F} D(x_i, y_i),$$

where for each $i \in F$, $D(x_i, y_i)$ is open. Since I is compact there exist $r < \omega$ and $G_p \in \mathfrak{F}$, for each $1 \le p \le r$, such that $\{G_p \mid 1 \le p \le r\}$ is disjoint and $F = \bigcup_{1 \le p \le r} G_p$, and for each $1 \le p \le r$ there exists $i_p \in F$ such that $G_p \subseteq D(x_{i_p}, y_{i_p})$. Let $G_o = I - F$ and let $x_{i_o} = y_{i_o} = x_{i_1}$. Since $\mathcal{B}$ is normal over $\mathfrak{F}$, there exist $x, y \in B$ such that for each $p \le r$,

$$G_p \subseteq E(x, x_{i_p}) \cap E(y, y_{i_p}).$$

It follows that $F = D(x, y)$ and (i) is established. Finally, (ii) follows from (i). ∎

Corollary 5.29 Suppose $\mathcal{B} \subseteq \Pi\langle \mathfrak{A}_i \mid i \in I\rangle$ is subdirect, where at most one $\mathfrak{A}_i$ is trivial, and $\mathfrak{F}$ is a field of subsets of I. Let $\mathfrak{T}(\mathfrak{F})$ be the topology on I with basis of clopen sets $\mathfrak{F}$. If $\mathcal{B}$ is normal over $\mathfrak{F}$ and I is compact then

$$\mathfrak{F} = \{E(x, y) \mid x, y \in B\} \cup \{D(x, y) \mid x, y \in B\}. \ ∎$$

Corollary 5.30 Suppose an algebra $\mathcal{B} \subseteq \Pi\langle \mathfrak{A}_i \mid i \in I\rangle$ is subdirect and $\mathfrak{F}$ is a field of subsets of I. Let $\mathfrak{T}(\mathfrak{F})$ be the topology on I with basis of clopen sets $\mathfrak{F}$, and let $J \subseteq I$ be a complete $\mathcal{B}$-transversal. If I

is compact and $\mathcal{B}$ is normal over $\mathfrak{F}$ then

(i) the quotient topology $\mathfrak{T}(\mathfrak{F})/\rho$ is compact and has a basis of clopen sets $\mathfrak{F}/\rho$, that is

$$\mathfrak{T}(\mathfrak{F})/\rho = \mathfrak{T}(\mathfrak{F}/\rho)$$

(ii) $\pi_J(\mathcal{B})$ is normal over $\mathfrak{F}/\rho$.

Proof: (i) Since $\rho: I \to J$ is a continuous retract, J is a compact space with the quotient topology $\mathfrak{T}(\mathfrak{F})/\rho$. To complete the proof of (i) we shall show that for every $j \in J$ and every open G, where $j \in G \subseteq J$, there exists F such that $j \in F \subseteq G$ and $\rho^{-1}(F) \in \mathfrak{F}$. In fact, let

$$E_o = \bigcap\{E(x, y) \mid x, y \in B\}$$

and consider $j \in J$ and open G, where $j \in G \subseteq J$. We distinguish two cases. Case 1: $G \cap E_o = \emptyset$. Then $j \in \rho^{-1}(G)$, where $\rho^{-1}(G) \cap E_o = \emptyset$. By Lemma 5.28 there exist $x, y \in B$ such that

$$j \in D(x, y) \subseteq \rho^{-1}(G).$$

Thus

$$j \in \rho(D(x, y)) = D(\pi_J(x), \pi_J(y)) \subseteq G.$$

By Lemma 2.37,

$$\rho^{-1}(D(\pi_J(x), \pi_J(y))) = D(x, y).$$

Case 2: $G \cap E_o \neq \emptyset$. Then $j \in \rho^{-1}(G)$, where $E_o \subseteq \rho^{-1}(G)$. Since E_o is closed, it is compact. Thus there exists $F \in \mathfrak{F}$ such that $j \in F$ and $E_o \subseteq F \subseteq \rho^{-1}(G)$. By Lemma 5.28, there exist $x, y \in B$ such that

$$j \in E(x, y) \subseteq \rho^{-1}(G).$$

Thus

$$j \in \rho(E(x, y)) = E(\pi_J(x), \pi_J(y)) \subseteq G$$

By Lemma 2.37,

$$\rho^{-1}(E(\pi_J(x), \pi_J(y))) = E(x, y).$$

This establishes (i).

(ii) By Lemma 4.4, $\mathcal{B}$ is $\mathfrak{T}(\mathfrak{F})$-global. Thus, by Theorem 2.39, $\pi_J(\mathcal{B})$ is $\mathfrak{T}(\mathfrak{F})/\rho$-global. It follows from part (i) and Lemma 4.3 that $\pi_J(\mathcal{B})$ patches over $\mathfrak{F}/\rho$. Finally, by Lemma 2.37, for every $x, y \in B$, $E(\pi_J(x), \pi_J(y)) \in \mathfrak{F}/\rho$. ∎

Theorem 5.31 Suppose an algebra $\mathcal{B} \subseteq \Pi\langle \mathfrak{A}_i \mid i \in I\rangle$ is a global subdirect product, where I is a zero-dimensional compact space and $E(x, y)$ is clopen for every $x, y \in B$. If $J \subseteq I$ is a complete $\mathcal{L}$-transversal then $\mathcal{L} \cong \pi_J(\mathcal{L})$, where $\pi_J(\mathcal{L}) \subseteq \Pi\langle \mathfrak{A}_j \mid j \in J\rangle$ is a Boolean subdirect product, at most one $\mathfrak{A}_j$ is trivial, and

$$\{E(\pi_J(x), \pi_J(y)) \mid x, y \in B\} \cup \{D(\pi_J(x), \pi_J(y)) \mid x, y \in B\}$$

is the field of all clopen sets of J.

<u>Proof</u>: Use Corollaries 5.29 and 5.30. ∎

Closure under the normal transform was first investigated in connection with so-called "Boolean powers" of (finite) algebras (See Gould and Gratzer [16]). In this context we should point out that we have changed the terminology. Our definition of a <u>Boolean subdirect power</u> of a structure $\mathfrak{A}$ is given at the beginning of this section. In the literature a <u>Boolean power</u> of $\mathfrak{A}$ is usually defined to be a structure of continuous functions $C_{\mathfrak{T}}(I, \mathfrak{A})$, where I is a Boolean space. To describe the relationship between these two notions, let $\mathfrak{A}$ be a non-trivial structure, give A the discrete topology and let $\mathfrak{T}$ be a topology on I.

<u>Lemma</u> 5.32 For every formula φ and every $x \in C_{\mathfrak{T}}(I, A)^n$, $[\![\varphi(x)]\!]$ is clopen.

<u>Proof</u>: Let $i \in [\![\varphi(x)]\!]$. Then

$$i \in \bigcap_{k<n} E(x_k, \overline{x_k(i)}) \subseteq [\![\varphi(x)]\!],$$

and therefore $[\![\varphi(x)]\!]$ is open. By the same argument, $[\![\neg\varphi(x)]\!]$ is open. ∎

Hence $C_{\mathfrak{T}}(I, \mathfrak{A})$ is a Boolean subdirect power of $\mathfrak{A}$ provided I is a Boolean space. Of course, the converse is not true in general, i.e. not every Boolean subdirect power of $\mathfrak{A}$ is a structure of continuous functions $C_{\mathfrak{T}}(I, \mathfrak{A})$, where I is a Boolean space. Moreover, in <u>our</u> definition of $C_{\mathfrak{T}}(I, \mathfrak{A})$ we do <u>not</u> require that I is a Boolean space (see Section 2). To see what this generalization amounts to, let $\mathfrak{A}$ be a non-trivial structure, give A the discrete topology and let $\mathfrak{T}$ be a topology on I. Let $\mathcal{G}$ be the field of clopen sets. Since $\mathfrak{A}$ is non-trivial,

$$\mathcal{G} = \{E(x, \bar{a}) \mid x \in C_{\mathfrak{T}}(I, A), a \in A\}$$

and $C_{\mathfrak{T}}(I, \mathfrak{A}) = C_{\mathfrak{T}(\mathcal{G})}(I, \mathfrak{A})$. Thus, without loss of generality, we may assume that I is zero-dimensional and $\mathfrak{T}$ is the equalizer topology induced by $C_{\mathfrak{T}}(I, \mathfrak{A})$.

Corollary 5.33 $C_{\mathfrak{T}}(I, \mathfrak{A})$ is closed under $\bar{n}$.

Proof: Since $C_{\mathfrak{T}}(I, \mathfrak{A})$ is global, by Lemma 5.32, $C_{\mathfrak{T}}(I, \mathfrak{A})$ is normal over the field of clopen sets. The assertion follows from Lemma 5.4. ∎

Lemma 5.34 If $\mathfrak{A}$ is a non-trivial algebra then $\mathcal{R}_0 = \breve{\mathcal{R}}_0 = \mathfrak{F}_0$ is the field of all clopen sets.

Proof: By Lemma 5.32, $\mathfrak{F}_0 \subseteq \mathcal{G} \subseteq \breve{\mathcal{R}}_0 \subseteq \mathfrak{F}_0$. Next, let $x, y \in C_{\mathfrak{T}}(I, A)$. Since $E(x, y) \in \breve{\mathcal{R}}_0$ and $\breve{\mathcal{R}}_0$ is a field, $D(x, y) \in \breve{\mathcal{R}}_0$ so that there exist $u, v \in C_{\mathfrak{T}}(I, A)$ such that $D(x, y) = E(u, v)$. Thus $E(x, y) = D(u, v) \in \mathcal{R}_0$. ∎

Corollary 5.35 If $\mathfrak{A}$ is a non-trivial algebra then the equalizer and hull-kernel topologies induced by $C_{\mathfrak{T}}(I, \mathfrak{A})$ are the same.

Proof: Use Corollaries 5.14 and 5.33, and Lemma 5.34. ∎

Of course, in general I is neither Hausdorff nor compact, that is $C_{\mathfrak{T}}(I, \mathfrak{A})$ is not a Boolean subdirect power of $\mathfrak{A}$. However, since $C_{\mathfrak{T}}(I, \mathfrak{A})$ is normal over the field of clopen sets, we can globalize $C_{\mathfrak{T}}(I, \mathfrak{A})$ to obtain a Boolean subdirect product, but in general we also obtain new factors (which are in $ISP_u \mathfrak{A}$). In case $\mathfrak{A}$ is finite, the globalization can be constructed more directly, as we shall quickly indicate. Let I be zero-dimensional with field $\mathcal{G}$ of clopen sets and let J be the set of prime filters of $\mathcal{G}$. Since $\mathfrak{A}$ is finite, for each $x \in C_{\mathfrak{T}}(I, A)$ and each $\mathcal{U} \in J$ there exists a unique $a \in A$ such that $E(x, \bar{a}) \in \mathcal{U}$. Define

$$g(x)(\mathcal{U}) = a$$

and verify that g is an isomorphism of $C_{\mathfrak{T}}(I, \mathfrak{A})$ onto $C_{\mathfrak{T}(\mathcal{G}^*)}(J, \mathfrak{A})$. Thus for finite $\mathfrak{A}$ we may as well assume that I is a Boolean space, as is commonly done in the literature.

Notice, if φ is an arbitrary formula and $x \in \Pi A_i$ then we have to be more careful defining $[\![\varphi(x)]\!]$. More specifically, if $\mathcal{B} \subseteq \Pi\langle \mathfrak{A}_i \mid i \in I\rangle$ and $x \in B$ then we define

$$[\![\varphi(x)]\!]^{\mathcal{B}} = \{i \in I \mid \pi_i(\mathcal{B}) \models \varphi[\pi_i(x)]\}$$

Of course, if φ is an open formula then no precautions are necessary.

Corollary 5.36 If $\mathcal{B} \subseteq C_{\mathfrak{T}}(I, \mathfrak{A})$ is $\mathfrak{T}$-global then for every open formula φ and every $x \in B^n$, $[\![\varphi(x)]\!]$ is clopen.

Proof: Use Lemma 5.32. ∎

Corollary 5.37 If $\mathcal{B} \subseteq C_{\mathfrak{T}}(I, \mathfrak{A})$ is $\mathfrak{T}$-global then $\mathcal{B}$ is closed under $\bar{n}$.

Proof: By Corollary 5.36, $\mathcal{B}$ is normal over the field of clopen sets. The assertion follows from Lemma 5.4. ∎

6. GLOBALLY REPRESENTABLE VARIETIES

Let $\mathcal{M}$ be a class of algebras. An algebra $\mathfrak{A}$ is called globally representable by $\mathcal{M}$ (Boolean representable by $\mathcal{M}$) if $\mathfrak{A}$ is isomorphic to a global subdirect product over a compact T_o-space (Boolean subdirect product), where each factor belongs to $\mathcal{M}$ or is trivial, and at most one factor is trivial. A variety $\mathcal{V}$ is called globally (Boolean) representable by $\mathcal{M}$ if every member of $\mathcal{V}$ is globally (Boolean) representable by $\mathcal{M}$. $\mathcal{V}$ is called a variety of lattice expansions if $\mathcal{V}$ is a variety and there are binary operation symbols $\wedge$ and $\vee$ such that for all $\mathfrak{A} \in \mathcal{V}$, $\langle A, \wedge^{\mathfrak{A}}, \vee^{\mathfrak{A}}\rangle$ is a lattice.

Lemma 6.1 Let $\mathcal{V}$ be a variety of lattice expansions and let $\mathcal{L}$ be the class of totally ordered members of $\mathcal{V}$. Then every subdirectly irreducible member of $V\mathcal{L}$ belongs to $\mathcal{L}$. Thus $V\mathcal{L} = IP_s\mathcal{L}$ and $\mathcal{L}$ is a universal class closed under homomorphism.

Proof: Since $\mathcal{V}$ is congruence distributive, by Jonsson [23], every subdirectly irreducible member of $V\mathcal{L}$ belongs to $HSP_s\mathcal{L} \subseteq \mathcal{L}$. ∎

Theorem 6.2 Let $\mathcal{V}$ be a congruence permutable variety of lattice expansions and let $\mathcal{L}$ be the class of totally ordered members of $\mathcal{V}$. Then $V\mathcal{L}$ is globally representable by $\mathcal{L}$.

Proof: Let $\mathfrak{A} \in V\mathcal{L}$ and let $\mathfrak{A} \cong \mathfrak{B} \subseteq \Pi\langle \mathfrak{A}_i \mid i \in I\rangle$ be subdirect, where $\{\theta_i^{\mathfrak{B}} \mid i \in I\}$ is the set of all completely meet irreducible congruences on $\mathfrak{B}$. By Lemma 6.1, $HSP_u\{\mathfrak{A}_i \mid i \in I\} \subseteq \mathcal{L}$ and the assertion follows from Theorem 4.42. ∎

We shall give three applications of Theorem 6.2 where the variety $V\mathcal{L}$ has attracted considerable attention in the literature.

An l-group (lattice ordered group) is an algebra $\mathfrak{B} = \langle B, \cdot, ^{-1}, e, \wedge, \vee\rangle$ where $\langle B, \cdot, ^{-1}, e\rangle$ is a group, $\langle B, \wedge, \vee\rangle$ is a lattice and for every $x, y, z \in B$

(i) if $x \leq y$ then $zx \leq zy$ and $xz \leq yz$.

As was observed by Stone [35], (i) can be replaced by either one of the following two equations:

(ii) $u(x \wedge y)v = uxv \wedge uyv$,

(iii) $u(x \vee y)v = uxv \vee uyv$,

so that the class of l-groups is a variety. An l-group is called a <u>vector group</u> if it is isomorphic to a subdirect product of totally ordered groups. By Lemma 6.1, the class of vector groups is a variety and an axiomatization is well-known (see Fuchs [14]):

(iv) $(x \wedge y^{-1}x^{-1}y) \vee e = e$.

The next result was first obtained by Ledbetter (written communication) using a different approach which we shall discuss later in this section.

<u>Corollary</u> 6.3 Every vector group is globally representable by totally ordered groups.

<u>Proof</u>: Use Theorem 6.2. ∎

An <u>l-ring</u> (<u>lattice ordered ring</u>) is an algebra $\mathcal{B} = \langle B, +, -, 0, \cdot, \wedge, \vee \rangle$ where $\langle B, +, -, 0, \cdot \rangle$ is a ring, $\langle B, \wedge, \vee \rangle$ is a lattice and for every $x, y, z \in B$,

(i) if $x \leq y$ then $x + z \leq y + z$;

(ii) if $x \leq y$ and $z \geq 0$ then $xz \leq yz$ and $zx \leq zy$.

As was observed by Birkhoff and Pierce [3], (ii) can be replaced by the equation

(iii) $(x \vee 0) \cdot (y \vee 0) \wedge 0 = 0$

so that the class of l-rings is also a variety. An l-ring is called a <u>function ring</u> (<u>f-ring</u>) if it is isomorphic to a subdirect product of totally ordered rings. By Lemma 6.1, the class of f-rings is a variety. This observation is again due to Birkhoff and Pierce [3] and we give an axiomatization which we obtained from Keimel [24]:

(iv) $(x \vee 0) \wedge (-x \vee 0) \cdot (y \vee 0) = 0 = (x \vee 0) \wedge (y \vee 0) \cdot (-x \vee 0)$.

The next result is due to Kennison [26].

<u>Corollary</u> 6.4 Every f-ring is globally representable by totally ordered rings.

<u>Proof</u>: Use Theorem 6.2 ∎

A <u>Heyting algebra</u> (<u>Brouwerian lattice</u>, <u>relatively pseudo-complemented lattice</u>) is an algebra $\mathcal{B} = \langle B, \wedge, \vee, 0, 1, \twoheadrightarrow \rangle$, where $\langle B, \wedge, \vee, 0, 1 \rangle$ is a lattice with zero and one and for every $x, y, z \in B$

(i) $x \wedge y \to y = 1$;

(ii) $x \wedge (x \to y) = x \wedge y$;

(iii) $x \wedge (x \wedge y \to z) = x \wedge (y \to z)$.

A Heyting algebra is called a relative Stone algebra if it is isomorphic to a subdirect product of totally ordered Heyting algebras. (See Balbes and Dwinger [2]). By Lemma 6.1, the class of relative Stone algebras is a variety and we give an axiomatization:

(iv) $(x \to y) \vee (y \to x) = 1$.

The next observation was again first made by Ledbetter (written communication).

Corollary 6.5 Every relative Stone algebra is globally representable by totally ordered Heyting algebras.

Proof: A Heyting algebra is congruence permutable and a Mal͡cev polynominal is due to Katriňák:

$$p(x, y, z) = (y \to x) \wedge (x \vee z) \wedge (y \to z)$$

Thus Theorem 6.2 applies again. ∎

A different approach with overlapping applications is due to Kennison and Ledbetter. We shall present Ledbetter's result (written communication) in a somewhat more general form.

Lemma 6.6 Suppose $\mathfrak{A}$ is arithmetical and $\mathfrak{M}$ is a universal class such that for every congruence θ on $\mathfrak{A}$,

(i) if θ is completely meet irreducible then $\mathfrak{A}/\theta \in \mathfrak{M}$;

(ii) if $\mathfrak{A}/\theta \in \mathfrak{M}$ then θ is meet irreducible.

Then $\mathfrak{A}$ is globally representable by $\mathfrak{M}$.

Proof: By (i) we obtain $\mathfrak{A} \cong \mathfrak{B} \subseteq \Pi\langle \mathfrak{A}_i \mid i \in I\rangle$ subdirect, where $\{\theta_i^{\mathfrak{B}} \mid i \in I\} = \{\psi \mid \mathfrak{B}/\psi \in \mathfrak{M}\}$. By Lemmas 3.13 and 4.40, $\mathfrak{B}$ strongly patches over the dual ring of hull-kernel closed sets. By Corollary 3.5, $\mathfrak{B}$ patches over the dual ring generated by $\{E(x, y) \mid x, y \in B\}$ and by Corollary 4.27, $\mathfrak{B}$ is global. ∎

Theorem 6.7 Let $\mathfrak{V}$ be an arithmetical variety and let $\mathfrak{M} \subseteq \mathfrak{V}$ be a universal class of subdirectly indecomposables containing all subdirectly irreducibles of $\mathfrak{V}$. Then $\mathfrak{V}$ is globally representable by $\mathfrak{M}$. ∎

Theorem 6.7 is applicable as soon as the universal class generated by the subdirectly irreducibles includes only subdirectly indecomposables. This condition is most readily verified in case the class of subdirectly indecomposables can be identified as a universal class itself. Ledbetter [29] has extracted from Kennison [26] sufficient algebraic conditions for the hypothesis of Theorem 6.7 to apply which often can be checked quite directly. However, in most cases the application of Theorem 6.7 is just as straight forward. For example, the subdirectly indecomposable vector groups are exactly the totally ordered groups and we obtain Corollary 6.3 (Ledbetter). Similarly, the subdirectly indecomposable f-rings are exactly the totally ordered rings and we obtain Corollary 6.4 (Kennison [26]). Finally, the subdirectly indecomposable relative Stone algebras are exactly the totally ordered Heyting algebras and we obtain Corollary 6.5 (Ledbetter). To give more applications of Theorem 6.7 we need some information about Heyting algebras.

<u>Lemma</u> 6.8 (i) A Heyting algebra is subdirectly irreducible if and only if it satisfies

$$\exists\, e\big[e \neq 1 \;\&\; \forall x\,[x = 1 \vee x \leq e]\big]$$

(ii) A Heyting algebra is subdirectly indecomposable if and only if it satisfies

$$\forall\, x\, \forall\, y\big[x \vee y = 1 \rightarrow [x = 1 \vee y = 1]\big]$$

(iii) The class of subdirectly indecomposable Heyting algebras is the universal class generated by the subdirectly irreducible Heyting algebras.

<u>Proof</u>: (i) is well-known. (ii) Suppose a Heyting algebra $\mathfrak{A}$ satisfies the given formula. Let F and G be filters on $\mathfrak{A}$, where $F \cap G = \{1\}$ but $F \neq \{1\}$. Suppose $1 \neq a \in F$ and $b \in G$. Then $a \vee b \in F \cap G = \{1\}$ so that $b = 1$. Thus $G = \{1\}$ and $\mathfrak{A}$ is subdirectly indecomposable. Conversely, suppose $\mathfrak{A}$ is subdirectly indecomposable and $a, b \in A$, where $a \vee b = 1$. Let $F = \{x \in A \mid x \geq a\}$ and $G = \{x \in A \mid x \geq b\}$. Then $F \cap G = \{1\}$ so that $a = 1$ or $b = 1$.

(iii) By (i) and (ii), the universal class generated by the subdirectly irreducible Heyting algebras is contained in the class of subdirectly indecomposable Heyting algebras. Conversely, let $\mathfrak{A}$ be a subdirectly indecomposable Heyting algebra and let $\mathfrak{B}$ be a Heyting algebra obtained from $\mathfrak{A}$ by inserting an element $e \notin A$ immediately below 1. Then $\mathfrak{B}$ is subdirectly irreducible and $\mathfrak{A} \subseteq \mathfrak{B}$. ∎

Corollary 6.9 Every Heyting algebra is globally representable by Heyting algebras in which 1 is join irreducible.

Proof: Use Theorem 6.7 and Lemma 6.8. ∎

A variety is called arithmetical if every member is arithmetical. We now have given four examples of arithmetical varieties which are globally representable by their subdirectly indecomposables: vector groups, function rings, relative Stone algebras and Heyting algebras.

Problem 6.10 Is every arithmetical variety globally representable by its subdirectly indecomposables?

We do not know whether (global) subdirect representation by subdirectly indecomposables is of much interest in general. In some cases we do obtain interesting results (Corollaries 6.3, 6.4 and 6.5) whereas in others the result appears to be rather dubious (Corollary 6.9). In contrast, global subdirect representation by subdirectly irreducibles becomes extremely interesting once it is considered as the non-trivial aspect of Birkhoff's Theorem. An algebra is called globally (Boolean) representable if it is globally (Boolean) representable by subdirectly irreducibles, and a variety is called globally (Boolean) representable in case every member is.

Problem 6.11 Which varieties are globally representable?

Problem 6.12 Which varieties are Boolean representable?

A solution of either one of these two problems would constitute a significant advance in the structure theory of varieties. We shall discuss some partial solutions which will illustrate the significance of these problems. Keimel and Werner [25] show that the variety generated by finitely many weakly independent quasi primal algebras is Boolean representable by subdirectly irreducibles, and Bullman-Fleming and Werner [4] generalize this result to discriminator varieties (See also Werner [39]). A variety $\mathcal{V}$ is called a discriminator variety if there exists a polynominal symbol p (that is a term in the corresponding equational language) such that for every subdirectly irreducible $\mathfrak{A} \in \mathcal{V}$, $p^{\mathfrak{A}} = n$. It is easy to see that a variety $\mathcal{V}$ is generated by finitely many weakly independent quasi primal algebras if and only if $\mathcal{V}$ is a discriminator variety containing finitely many subdirectly irreducibles (Werner [39]).

Lemma 6.13 Suppose $\{\mathfrak{A}_i \mid i \in I\}$ is a family of algebras and for each $i \in I$, $f_i : A_i^r \to A_i$. Then every subalgebra of every direct product of members of $\{\mathfrak{A}_i \mid i \in I\}$ is closed under the induced pointwise operation $\bar{f}$ if and only if there exists a polynominal symbol p such that for every $i \in I$, $f_i = p^{\mathfrak{A}_i}$.

Proof: Suppose every subalgebra of every direct product of members of $\{\mathfrak{A}_i \mid i \in I\}$ is closed under the induced pointwise operation $\bar{f}$. We may assume that $\{\mathfrak{A}_i \mid i \in I\}$ is disjoint. Consider

$$\mathfrak{A} = \Pi \langle \mathfrak{A}_a \mid a \in \bigcup_{i \in I} A_i^r \rangle,$$

where $\mathfrak{A}_a = \mathfrak{A}_i$ for $a \in A_i^r$. Let $\mathfrak{B}$ be the subalgebra of $\mathfrak{A}$ generated by $\{x_p \mid p < r\}$, where

$$x_p(a) = a_p \quad \text{for } a \in A_i^r$$

By hypothesis, $\bar{f}(x_o, \ldots, x_{r-1}) \in B$ and therefore there exists a polynominal symbol p such that

$$p^{\mathfrak{A}}(x_o, \ldots, x_{r-1}) = \bar{f}(x_o, \ldots, x_{r-1}).$$

Thus for each $i \in I$ and $a \in A_i^r$,

$$f_i(a) = \bar{f}(x_o, \ldots, x_{r-1})(a)$$
$$= p^{\mathfrak{A}}(x_o, \ldots, x_{r-1})(a) = p^{\mathfrak{A}_i}(a).$$

This proves one direction of the assertion and the other is trivial. ∎

Theorem 6.14 If $\mathcal{V}$ is a variety then the following are equivalent:

(i) $\mathcal{V}$ is a discriminator variety.

(ii) Every subdirect product of subdirectly irreducible members of $\mathcal{V}$ is closed under $\bar{n}$.

(iii) $\mathcal{V}$ is congruence distributive and every member of $\mathcal{V}$ is quasi regular.

Proof: The equivalence of (i) and (ii) follows from Lemma 6.13. Next, assume (ii). By (i), $\mathcal{V}$ is congruence distributive and by Lemma 5.21, every member of $\mathcal{V}$ is quasi regular. Conversely, (ii) follows from (iii) by Lemmas 5.19 and 5.21. ∎

Theorem 6.15 (Bulman-Fleming and Werner) Every discriminator variety is Boolean representable.

Proof: Use Theorems 5.24 and 6.14. ∎

A variety is called semi simple if every subdirectly irreducible member is simple.

Corollary 6.16 Suppose $\mathcal{V}$ is a congruence distributive semi simple variety. Then $\mathcal{V}$ is Boolean representable if and only if $\mathcal{V}$ is a discriminator variety.

Proof: Use Theorem 5.24 and Theorems 6.14 and 6.15. ∎

It follows at once from Corollary 6.16 that the variety of distributive lattices (or, more generally, a dual discriminator variety which is not congruence permutable - see Fried and Pixley [15]) is not Boolean representable.

Problem 6.17 Is the variety of distributive lattices (or, more generally, a dual discriminator variety) globally representable?

Returning to arithmetical varieties we have examples which are globally representable but not discriminator varieties:

Corollary 6.18 The variety generated by a finite Heyting algebra is globally representable.

Proof: Let $\mathfrak{A}$ be a finite Heyting algebra. By Jonsson [23], every subdirectly irreducible member of $V\mathfrak{A}$ belongs to $HS\mathfrak{A}$, and therefore there are (up to isomorphism) only finitely many finite ones. Since a subalgebra of a finite subdirectly irreducible Heyting algebra is subdirectly irreducible, the class of subdirectly irreducible members of $V\mathfrak{A}$ is universal. Now apply Theorem 6.7. ∎

Problem 6.19 Is an arithmetical variety globally representable? More specifically, is the class of f-rings globally representable?

In another direction we make the following observation:

Theorem 6.20 The variety generated by a finite abelian group in Boolean representable.

Proof: Let $\mathfrak{A}$ be a finite abelian group and let $\mathcal{B} \in V\mathfrak{A}$. Then $\mathcal{B}$

is isomorphic to a direct sum of subgroups of $\mathfrak{A}$ with prime power order. The assertion follows from Theorem 5.26. ∎

Problem 6.21 Is the variety generated by a para primal algebra globally (Boolean) representable?

This assortment of facts and questions should amply demonstrate that Problems 6.11 and 6.12 are both difficult and interesting.

7. GLOBAL SUBDIRECT REPRESENTATION OF RINGS

In this section we shall apply our method to obtain some sheaf representation theorems for rings. There is a vast literature on the subject and it cannot be the purpose of this paper to rewrite the sheaf representation theory of rings. Therefore we have chosen a few representative examples which are considered significant results in the subject matter and which are characteritic for the methods employed. It turns out that we only need a minute amount of elementary ring theory to obtain these results directly from our theorems. However, as we would like to convince algebraists of the considerable advantages of our methods over the conventional approach, we shall give a rather expansive exposition of the background of our applications to ring theory.

Throughout this section we shall consider rings $\mathcal{B} = \langle B, +, -, 0, \cdot\rangle$. If $\mathcal{B}$ is a ring and ψ is a congruence on $\mathcal{B}$ then

$$M(\psi) = \{x \in B \mid x\psi 0\}$$

is an ideal of $\mathcal{B}$ and M is an isomorphism from the congruence lattice of $\mathcal{B}$ onto the lattice of ideals of $\mathcal{B}$. Via this isomorphism all facts about congruences can be coverted into facts about ideals and vice versa. To obtain a subdirect representation of $\mathcal{B}$ inducing a hull-kernel topology we have to consider the space of irreducible congruences (ideals) on $\mathcal{B}$. Accordingly, let

$$\begin{array}{ll} \text{Max}\ \mathcal{B} & (\text{Max}\ \Theta(\mathcal{B})) \\ \text{Ptv}\ \mathcal{B} & (\text{Ptv}\ \Theta(\mathcal{B})) \\ \text{Prm}\ \mathcal{B} & (\text{Prm}\ \Theta(\mathcal{B})) \\ \text{Irrd}\ \mathcal{B} & (\text{Irrd}\ \Theta(\mathcal{B})) \end{array}$$

be the spaces of <u>maximal</u>, <u>primitive</u>, <u>prime</u> and <u>irreducible</u> ideals (congruences) of $\mathcal{B}$ respectively. The following facts are well-known (c.f. Jacobson [22] or Lambek [28]).

<u>Lemma</u> 7.1 If $\mathcal{B}$ is a ring then

(i) $\text{Ptv}\ \mathcal{B} \subseteq \text{Prm}\ \mathcal{B} \subseteq \text{Irrd}\ \mathcal{B}$.

(ii) $\text{Max}\ \mathcal{B} \subseteq \text{Prm}\ \mathcal{B}$ in case $\mathcal{B}$ has an identity.

(iii) $\text{Max}\ \mathcal{B} = \text{Ptv}\ \mathcal{B}$ in case $\mathcal{B}$ is commutative with identity.

(iv) $\text{Max}\ \mathcal{B} = \text{Ptv}\ \mathcal{B} = \text{Prm}\ \mathcal{B}$ in case $\mathcal{B}$ is commutative and Artinian with identity. ∎

By Corollary 3.8, $\mathcal{L}$ induces a hull-kernel topology on Irrd $\Theta(\mathcal{L})$ and its subspaces. Notice that this topology is T_0. Prm $\mathcal{L}$ is usually called the (prime) spectrum of $\mathcal{L}$, and is the largest space of irreducible ideals commonly considered in the literature. To obtain a subdirect representation of $\mathcal{L}$ inducing a hull-kernel topology we use Corollary 3.10 and define $\mathcal{L}$ to be semi-irreducible (semi-prime, semi-simple and strongly semi-simple) in case the intersection of Irrd $\Theta(\mathcal{L})$ (Prm $\Theta(\mathcal{L})$, Ptv $\Theta(\mathcal{L})$, Max $\Theta(\mathcal{L})$) is $\mathbb{O}^{\mathcal{L}}$. Curiously enough, the class of semi-irreducible rings has not attracted much attention in ring theory. The ring of integers with zero-multipication is semi-irreducible but not semi-prime.

Next we extract from Hofmann [20] the basic algebraic fact that produces patching over the dual ring of hull-kernel open sets:

Lemma 7.2 Let $\mathcal{L} \subseteq \Pi\langle \mathfrak{A}_i \mid i \in I\rangle$ be a ring with identity, where Max $\Theta(\mathcal{L}) \subseteq \{\theta_i^{\mathcal{L}} \mid i \in I\} \subseteq$ Irrd $\Theta(\mathcal{L})$. Then $\mathcal{L}$ induces a hull-kernel topology on I and patches over the dual ring $\mathcal{R}$ of hull-kernel open sets.

Proof: By Corollary 3.8, $\mathcal{L}$ induces a hull-kernel topology on I. Next, let $r < \omega$ and $F_p \in \mathcal{R}$, $x_p \in B$ for each $p < r$, where $I = \bigcup_{p<r} F_p$ and $F_p \cap F_q \subseteq E(x_p, x_q)$ whenever $p, q < r$. Since $1 \in B$, every congruence on $\mathcal{L}$ is contained in a maximal congruence so that Lemma 3.29 applies:

$$\mathbb{1}^{\mathcal{L}} = \bigvee_{p<r} \theta^{\mathcal{L}}_{I-F_p}$$

In terms of ideals,

$$B = \sum_{p<r} M(\theta^{\mathcal{L}}_{I-F_p}).$$

Thus $1 = \sum_{p<r} b_p$, where $b_p \in M(\theta^{\mathcal{L}}_{I-F_p})$ for each $p < r$. That is, $D(b_p, 0) \subseteq F_p$ for each $p < r$. Now let $y = \sum_{p<r} x_p b_p$ and verify that $F_p \subseteq E(y, x_p)$ for all $p < r$. ∎

Theorem 7.3 Let $\mathcal{L} \subseteq \Pi\langle \mathfrak{A}_i \mid i \in I\rangle$ be a semi-irreducible (semi-prime) ring with identity, where $\{\theta_i^{\mathcal{L}} \mid i \in I\}$ is the set of all irreducible (prime) congruences on $\mathcal{L}$. Then $\mathcal{L}$ induces a compact T_0 hull-kernel topology on I and patches over the dual ring of hull-kernel open sets.

Proof: By Lemma 7.1, Max $\Theta(\mathcal{L}) \subseteq \{\theta_i^{\mathcal{L}} \mid i \in I\}$. The assertion follows from Lemmas 3.28 and 7.2. ∎

Theorem 7.3 in this form apparently has not been noticed in the literature and deserves commenting. Call $\mathcal{B} \subseteq \Pi\langle \mathfrak{A}_i \mid i \in I\rangle$ <u>dually normal</u> over the ring $\mathcal{R}$ if

(i) $\mathcal{B}$ patches over $\breve{\mathcal{R}}$;

(ii) for every atomic formula φ and every $x \in B^n$, $[\![\varphi(x)]\!] \in \mathcal{R}$.

Then in Theorem 7.3 $\mathcal{B}$ is dually normal over the ring of hull-kernel closed sets. Now if normalcy is a good notion of "tightness" then dual normalcy ought to be too. The trouble is that equalizers are not open but closed. Therefore we cannot utilize compactness of the space to extend (finite) patching to unrestricted patching (compare Lemma 4.4), and we are stuck with a <u>finitary</u> notion of tightness. Nevertheless, Theorem 7.3 ought to be considered a rather "nice" subdirect representation of the ring $\mathcal{B}$ with fairly "nice" factors (such as prime rings). If we are not satisfied with this condition of "tightness" then we have to shove $\mathcal{B}$ into the globalization mill: Out comes a global subdirect representation of $\mathcal{B}$, where in general we know practically nothing about the factors. This is what Hofmann [20] does to obtain his celebrated Theorem 1.17. We obtain a somewhat more general version:

<u>Theorem</u> 7.4 Let $\mathcal{B} \subseteq \Pi\langle \mathfrak{A}_i \mid i \in I\rangle$ be a semi irreducible (semi-prime) ring with identity, where $\{\theta_i^{\mathcal{B}} \mid i \in I\}$ is the set of all irreducible (prime) congruences on $\mathcal{B}$. Then $\mathcal{B}$ is isomorphic to a global subdirect product

$$\mathcal{B} \cong g(\mathcal{B}) \subseteq \Pi\langle \mathcal{B}/h(i) \mid i \in I\rangle,$$

where I is a compact T_o-space and

$$x \equiv y(h(i)) \text{ iff } i \in E(x, y)^o.$$

<u>Proof</u>: Use Remarks 4.20 and Theorem 7.3. ∎

It is easy to verify that in the case of semi-prime rings Theorem 7.4 is Hofmann's result. Notice that the sum total of ring theory involved is the patching from Lemma 7.2. Moreover, <u>in general</u> ring theory does not appear to yield more information about the new factors $\mathcal{B}/h(i)$ than universal algebra. Wolf [41] makes an attempt to explain sheaf representation in the setting of universal algebra and comes up with his version of Theorem 4.38. He then notices that this explanation does not apply to Theorem 7.4. The reason is quite simple. Wolf requires strong patching, and in Lemma 7.2 we do not in general obtain strong patching over the dual ring of hull-kernel open sets.

To obtain more information about the new factors $\mathcal{B}/h(i)$ in Theorem 7.4, ring theorists impose additional restrictions on the ring $\mathcal{B}$. In many cases these restrictions are so severe that patching can be obtained much more directly from closure under the normal transform than from Lemma 7.2. We shall first look at a case where we need Lemma 7.2 for patching.

Dauns and Hofmann [12] call a ring <u>weakly biregular</u> if for any two distinct prime ideals M and N of $\mathcal{B}$ there exists a central idempotent $e \in B$ with $e \in M$ and $e \notin N$. In particular, this makes the prime ideal space T_1. We again extract the relevant algebraic fact from Hofmann [20]:

<u>Lemma</u> 7.5 Let $\mathcal{B}$ be a semi-prime ring with identity. If $\mathcal{B}$ is weakly biregular then $\mathrm{Max}\,\mathcal{B} = \mathrm{Prm}\,\mathcal{B}$ is a Boolean space.

<u>Proof</u>: Let $\mathcal{B} \subseteq \Pi\langle \mathfrak{A}_i \mid i \in I\rangle$ be subdirect, where $\{\theta_i^{\mathcal{B}} \mid i \in I\} = \mathrm{Prm}\,\Theta(\mathcal{B})$. By Lemma 7.1, $\mathrm{Max}\,\Theta(\mathcal{B}) \subseteq \mathrm{Prm}\,\Theta(\mathcal{B})$. Let $E(\mathcal{B})$ be the set of central idempotents of $\mathcal{B}$. Now let $\theta_i^{\mathcal{B}} \in \mathrm{Prm}\,\Theta(\mathcal{B})$. Since $1 \in B$, there exists $\theta_j^{\mathcal{B}} \in \mathrm{Max}\,\Theta(\mathcal{B})$ where $\theta_i^{\mathcal{B}} \subseteq \theta_j^{\mathcal{B}}$. If $\theta_i^{\mathcal{B}} \neq \theta_j^{\mathcal{B}}$ then there exists $e \in E(\mathcal{B})$ such that $i \in E(e, 0)$ and $j \notin E(e, 0)$, contradicting $j \in \overline{\{i\}} \subseteq E(e, 0)$. Thus $\theta_i^{\mathcal{B}} = \theta_j^{\mathcal{B}}$ and $\mathrm{Max}\,\Theta(\mathcal{B}) = \mathrm{Prm}\,\Theta(\mathcal{B})$.

By Lemma 3.28, I is compact. If $e \in E(\mathcal{B})$ then $1 - e \in E(\mathcal{B})$ and $e(1 - e) = 0$. Thus every prime ideal contains exactly one of e and $1 - e$. Thus for each $e \in E(\mathcal{B})$, $D(e, 0) = E(1 - e, 0)$ is clopen. It remains to show that

$$\{D(e, 0) \mid e \in E(\mathcal{B})\}$$

is a subbasis for the hull-kernel topology. Let $x \in B$ and $i \in D(x, 0)$. For each $j \in E(x, 0)$ there is $e_j \in E(\mathcal{B})$ where $j \in E(e_j, 0)$ and $i \notin E(e_j, 0)$, because $\mathcal{B}$ is weakly biregular. By compactness there exists finite $J \subseteq E(x, 0)$ such that

$$E(x, 0) \subseteq \bigcup_{j\in J} E(e_j, 0).$$

Thus

$$i \in \bigcap_{j\in J} D(e_j, 0) \subseteq D(x, 0). \blacksquare$$

Hofmann [20] calls a ring <u>local</u> if it has a maximal ideal containing every proper ideal.

<u>Theorem</u> 7.6 (Dauns and Hofmann [12]) Every weakly biregular semi-prime ring with identity is isomorphic to a global subdirect product of local rings over a Boolean space.

Proof: Use Lemma 7.5 to obtain $\mathcal{L} \subseteq \Pi\langle \mathfrak{A}_i \mid i \in I\rangle$ subdirect, where $\{\theta_i^{\mathcal{L}} \mid i \in I\} = \text{Max } \Theta(\mathcal{L}) = \text{Prm } \Theta(\mathcal{L})$ is a Boolean space. By Theorem 7.4, $\mathcal{L}$ is isomorphic to a global subdirect product

$$\mathcal{L} \cong g(\mathcal{L}) \subseteq \langle \mathcal{L}/h(i) \mid i \in I\rangle,$$

where $x \equiv y(h(i))$ if and only if $i \in E(x, y)^{\circ}$. Now $\theta_i^{\mathcal{L}}$ is the only maximal congruence containing $\equiv(h(i))$, so $\mathcal{L}/h(i)$ is a local ring. ∎

Notice that Theorem 7.6 does not yield a Boolean subdirect representation. Moreover, to know that the factors are local rings does not amount to that much either. Apparently most global subdirect representations which really give some information about the factors are Boolean. Recall from Corollary 5.8 that a Boolean subdirect product is closed under the normal transform. Frequently it is fairly easy to establish directly that some "natural" subdirect representation is closed under the normal transform. Subsequently we can apply the machinery of Section 5 to obtain the desired Boolean subdirect representation. One of the most striking applications of this procedure yields the celebrated results of Dauns and Hofmann [11] on biregular and commutative regular rings. A ring $\mathcal{L}$ is called biregular if every principal ideal is generated by a central idempotent. $\mathcal{L}$ is called regular if for each $x \in B$ there exists $y \in B$ such that $xyx = x$. Notice that a commutative regular ring is biregular. Biregular rings were first introduced by Arens and Kaplansky [1] who built on Jacobson's [21] original work on the hull-kernel topology to obtain sheaf representations for certain biregular rings. Dauns and Hofmann [11] extended their results to represent every biregular (commutative regular) ring as the structure of all global sections with compact support in a sheaf of simple rings with identity (fields) over a locally compact, zero-dimensional Hausdorff space. It is again striking to realize that only a minute amount of ring theory is required beyond universal algebra, and it is historically interesting to notice that the pertinent ring theory is explicitly stated and proved in Arens and Kaplansky [1]. In fact, Dauns and Hofmann's theorem follows directly from our results of Section 5 and a few elementary facts of ring theory which we extract from Arens and Kaplansky [1]:

Lemma 7.7 If $\mathcal{L}$ is a biregular ring then

(i) $\mathcal{L}$ is semi-simple;

(ii) every homomorphic image of $\mathcal{L}$ is biregular;

(iii) $\text{Ptv } \mathcal{L} = \text{Max } \mathcal{L}$.

Proof: (i) follows from the fact that the Jacobson radical N of $\mathcal{L}$ can contain no non-zero idempotent. Indeed, if $e^2 = e$ and $e \in N$ then $e + z - ex = 0$ for some $x \in B$. Multiplying on the left by e yields $e = 0$. (ii) is immediate. To prove (iii) we use (ii) and argue that if $\mathcal{B}$ is primitive then it is simple. Let $0 \neq x \in B$. We shall show that $\langle x \rangle = B$, where $\langle x \rangle$ is the ideal generated by x. Since $\mathcal{B}$ is biregular, $\langle x \rangle = \langle e \rangle$ for some central idempotent $e \in B$. Now let M be a faithful irreducible $\mathcal{B}$-module. Since $e \neq 0$, $Me \neq \{0\}$, so that $Me = M$. Let $y \in B$. If $m \in M$ and $0 \neq me \in M$ then $me(ey - y) = 0$, so that $ey = y \in \langle x \rangle$. ∎

By Lemma 7.7 a biregular ring $\mathcal{B}$ is a subdirect product of simple rings:

$$\mathcal{B} \subseteq \Pi\langle \mathfrak{A}_i \mid i \in I \rangle, \text{ where } \{\theta_i^{\mathcal{B}} \mid i \in I\} = \text{Max}\,\Theta(\mathcal{B}).$$

Since each factor $\mathfrak{A}_i$ is a simple biregular ring, each factor has an identity. This is the "natural" subdirect representation to start with. To obtain the relevant properties of this representation we again refer to Arens and Kaplansky [1]:

Lemma 7.8 $\mathcal{B} \subseteq \Pi\langle \mathfrak{A}_i \mid i \in I\rangle$ is fully expanded.

Proof: We shall show that every proper ideal of $\mathcal{B}$ is an intersection of maximal ideals. Let M be an ideal of $\mathcal{B}$, $x \in M$, and let N be an ideal that is maximal with respect to containing M but not x. Then $\mathcal{B}/N$ is subdirectly irreducible and, by Lemma 7.7, is biregular. Thus $\mathcal{B}/N$ is a subdirect product of simple rings and therefore is simple. We conclude that N is maximal. ∎

Lemma 7.9 $\mathcal{B} \subseteq \Pi\langle \mathfrak{A}_i \mid i \in I$ is closed under $\bar{n}$.

Proof: Let $x, y, u, v \in B$. Then the central idempotent e generating the ideal $\langle x - y \rangle$ is the characteristic function of $D(x, y)$ and

$$\bar{n}(x, y, u, v) = u + (v - u)\,e \in B. \;∎$$

Well, that's it. The rest is universal algebra.

Theorem 7.10 (Dauns and Hofmann [11]) Let $\mathcal{B}$ be a non-trivial biregular (commutative regular) ring with identity. If $\mathcal{B} \subseteq \Pi\langle \mathfrak{A}_i \mid i \in I\rangle$ is subdirect where $\{\theta_i^{\mathcal{B}} \mid i \in I\} = \text{Max}\,\Theta(\mathcal{B})$, then $\mathcal{B}$ is a Boolean subdirect product of non-trivial simple rings with identity (fields).

Proof: By Lemmas 7.8 and 7.9, $\mathcal{B} \subseteq \Pi\langle \mathfrak{A}_i \mid i \in I\rangle$ is fully expanded and closed under $\bar{n}$. Since $I = D(0, 1)$, Theorem 5.17 (ii) applies. ∎

If $\mathcal{B}$ does not have an identity then by Lemma 5.16, I is not compact. In this case we have three options. (Recall that $D(x, 0)$ is called the <u>support</u> of $x \in \Pi A_i$).

<u>Theorem</u> 7.11 Let $\mathcal{B}$ be a non-trivial biregular (commutative regular) ring without identity. If $\mathcal{B} \subseteq \Pi\langle \mathfrak{A}_i \mid i \in I\rangle$ is subdirect, where $\{\theta_i^{\mathcal{B}} \mid i \in I\} = \text{Max}\,\Theta(\mathcal{B})$ then

(i) (Dauns and Hofmann [11]) I is a locally compact, zero-dimensional Hausdorff space, and $\mathcal{B}$ is the ring of all global sections with compact support in $\Gamma(\mathcal{B}, \mathfrak{T})$, where $\mathfrak{T}$ is the hull-kernel topology induced by $\mathcal{B}$ and $\Gamma(\mathcal{B}, \mathfrak{T})$ is a global subdirect product of non-trivial simple rings with identity (fields);

(ii) $\mathcal{B}$ is isomorphic to a Boolean subdirect product of non-trivial simple rings with identity (fields) and one trivial ring;

(iii) $\mathcal{B}$ is a global subdirect product of non-trivial rings with identity (fields) over a compact T_1-space.

<u>Proof</u>: Use Lemmas 7.8 and 7.9. Then (i) follows from Theorem 5.18, (ii) follows from Theorem 5.17 (iii), and (iii) follows from Theorem 5.17 (i) and Lemma 5.15. ∎

Dauns and Hofmann obtain converses to their results in which the identity plays a rather camouflaged role. The point is rather interesting and we shall quickly explain what is involved. If $\mathcal{B} \subseteq \Pi\langle \mathfrak{A}_i \mid i \in I\rangle$ is a global subdirect product of non-trivial simple rings with identity over a locally compact, zero-dimensional Hausdorff space <u>and if</u> $1 \in B$, <u>where</u> $1(i)$ <u>is the identity of</u> $\mathfrak{A}_i$ <u>for each</u> $i \in I$, then the ring $\mathcal{B}_0$ of all members of $\mathcal{B}$ with compact support is a biregular ring. Now it is easy to see that in Theorem 7.11 (i), $1 \in \Gamma(B, \mathfrak{T})$ even in case $\mathcal{B}$ does not have an identity. Indeed, let $y \in B$ and let $e \in B$ be the central idempotent generating the same ideal as y. Then by Corollary 5.14,

$$E(y, 1) = E(y, e) \cap E(e, 1) = E(y, e) \cap D(e, 0) \in \mathfrak{T}$$

and we conclude that $1 \in \Gamma(B, \mathfrak{T})$. In particular, if $\mathcal{B}$ does not have an identity then the global closure of $\mathcal{B}$ from Theorem 7.11 (i) is not isomorphic to the globalization of $\mathcal{B}$ from Theorem 7.11 (ii).

We conclude our discussion of biregular rings with a simple observation.

Corollary 7.12 A biregular ring is arithmetical.

Proof: Use Lemma 7.8 and Theorem 3.34. ∎

Wolf [41] uses this observation and his version of Theorem 4.38 to obtain Dauns and Hofmann's results. Since this is an extremely circuitous approach we shall not indulge in its persuit. However, the so-called Pierce sheaf yields genuinely new sheaf representations of rings with identity (Pierce [31]). Let $\mathcal{B}$ be a ring with identity and let $E(\mathcal{B})$ be the set of central idempotents of $\mathcal{B}$. $E(\mathcal{B})$ is a Boolean algebra, where

$$e \wedge f = e \cdot f$$
$$e \wedge f = e + f - e \cdot f$$
$$\bar{e} = 1 - e.$$

For each $e \in E(\mathcal{B})$, let ψ_e be the congruence of $\mathcal{B}$ determined by the principal ideal eB, that is for each $x, y \in B$,

$$x \psi_e y \quad \text{iff} \quad x(1 - e) = y(1 - e).$$

Then ψ is an isomorphism of $E(\mathcal{B})$ onto a sublattice of the congruence lattice of $\mathcal{B}$ which is a Boolean algebra containing $\mathbf{0}^{\mathcal{B}}$ and $\mathbf{1}^{\mathcal{B}}$, and we can apply Theorem 4.34. (Of course, Comer's Theorem was extracted from the Pierce construction!). The resulting sheaf is generally known as the Pierce sheaf of $\mathcal{B}$, and we shall call the corresponding global subdirect product the Pierce representation of $\mathcal{B}$. In this case much effort has gone into the investigation of the factors of the Pierce representation, and we shall relate one of the main results to our investigation. A ring $\mathcal{B}$ is called a Baer ring if for every $x \in B$ the set

$$x^{\perp} = \{y \in B \mid xy = 0 = yx\}$$

is the principal ideal generated by some central idempotent. Notice that the generator of $0^{\perp}$ in a Baer ring is an identity. A ring is called a domain if it contains no zero-divisors.

Theorem 7.13 If $\mathcal{B}$ is a ring with identity then the following are equivalent:

(i) $\mathcal{B}$ is a Baer ring.

(ii) The Pierce shaef of $\mathcal{B}$ is a Hausdorff sheaf of domains over a Boolean space.

(iii) The Pierce representation of $\mathcal{B}$ is closed under $\bar{n}$ and its factors are domains.

Proof: The equivalence of (i) and (ii) is due to Hofmann [20]. By Corollary 2.33, (ii) is equivalent to

(ii') The Pierce representation of $\mathcal{B}$ is a Boolean subdirect product of domains.

Finally, since in the Pierce representation of $\mathcal{B}$, $I = D(0, 1)$, the equivalence of (ii') and (iii) follows from Lemma 5.7. ∎

Since the class of domains is universal, Corollary 4.27 suggests a characterization of global subdirect products of domains and Kennison [26] succeeded for integral domains (that is commutative domains with identity). Since a commutative ring with identity is isomorphic to a subdirect product of integral domains if and only if it has no non-zero nilpotent elements, the issue in Corollary 4.27 is patching. For each positive $n < \omega$, Kennison defines the n-th condition for domain representability DR_n to be the formula

$$[y \prod_{i<n} (y - x_i^2) = 0 \;\&\; \bigwedge_{i<j<n} z_i z_j = x_i x_j y] \rightarrow \exists w [w^2 = y \;\&\; \bigwedge_{i<n} x_i w = z_i]$$

and then establishes the following result which we state without proof:

Lemma 7.14 Suppose $\mathcal{B}$ is a commutative ring with identity and $\mathcal{B} \subseteq \Pi\langle \mathfrak{A}_i \mid i \in I\rangle$ is subdirect, where $\{\theta_i^{\mathcal{B}} \mid i \in I\} = \mathrm{Prm}\,\Theta(\mathcal{B})$. Then $\mathcal{B}$ patches over the dual ring generated by $\{E(x, y) \mid x, y \in B\}$ if and only if $\mathcal{B}$ satisfies DR_n for all positive $n < \omega$. ∎

Corollary 7.15 (Kennison [26]) A commutative ring with identity is isomorphic to a global subdirect product of integral domains if and only if it has no non-zero nilpotent elements and satisfies DR_n for all positive $n < \omega$.

Proof: Use Corollary 4.27 and Lemma 7.14. ∎

As indicated before, we suspect that globalization does not always improve a subdirect representation. Although we shall defer to the judgment of ring theorists, we feel tempted to dispute the superiority of Theorem 7.4 over Theorem 7.3. To further illustrate this point we shall discuss an example where a perfectly beautiful subdirect representation is globalized into a rather strange sheaf representation. To begin with, let I be a topological space and let $C(I, R)$ be the ring of continuous functions from I into the set R of real numbers. This "subdirect representation" of $C(I, R)$ appears to us as "nice" as can be. But now we make a few simple observations:

Lemma 7.16 $C(I, R)$ induces a hull-kernel topology on I which is contained in the topology of I. Moreover, if I is a metric space then the topology of I is exactly the hull-kernel topology induced by $C(I, R)$.

Proof: If $x, y \in C(I, R)$ then $D(x, 0) \cap D(x, 0) = D(xy, 0)$, so that

$$\{D(x, y) \mid x, y \in C(I, R)\}$$

is a basis for a topology on I. By Corollary 3.5, this is the hull-kernel topology induced by $C(I, R)$. Since R is Hausdorff, $D(x, y)$ is open for every $x, y \in C(I, R)$. Conversely, if I is a metric space then each basic open set is of the form $D(x, 0)$ for some $x \in C(I, R)$. ∎

Our next observation is immediate.

Lemma 7.17 $C(I, R)$ patches over the dual ring of open subsets of I. ∎

Hence we can globalize. First define the ideals

$$M(i) = \{x \in C(I, R) \mid x(i) = 0\}$$
$$H(i) = \{x \in C(I, R) \mid i \in E(x, 0)^{o}\}.$$

Theorem 7.18 If I is compact then $C(I, R)$ is isomorphic to a global subdirect product of local rings over I, where the quotient of each factor by its maximal ideal is isomorphic to the reals.

Proof: By Remarks 4.20 and Lemma 4.8, $C(I, R)$ is isomorphic to a global subdirect product of $\{C(I, R)/H(i) \mid i \in I\}$. Since I is compact, $\{M(i) \mid i \in I\}$ is the set of maximal ideals of $C(I, R)$. Thus $M(i)$ is the unique maximal ideal containing $H(i)$, so $C(I, R)/H(i)$ is local and its quotient by its maximal ideal is isomorphic to $C(I, R)/M(i)$. ∎

The ideal $H(i)$ is usually called the germinal ideal at i and the sheaf of Theorem 7.18 is called the sheaf of germs of continuous functions. In our view Theorem 7.18 is a perfect illustration of what appears to happen frequently in the process of globalization. The factors of the original "subdirect representation" $C(I, R)$ are the real numbers, and the representation itself must be one of the best understood pieces of mathematics. Nevertheless, equalizers are closed rather than open. Globalizing we turn the situation around at the expense of blowing up the reals into local rings which have a huge maximal ideal with a "nice" quotient ring, namely the reals. We fail to see the improvement accomplished by this process.

We should like to pursue this point a little further. In the proof of

Theorem 7.18 we needed the fact that every maximal ideal of C(I, R) is of the form M(i) for some $i \in I$. This fact depends on a property of the real numbers which is not shared by the complex numbers, namely $\sum_{i<n} x_i^2 \neq 0$ whenever $x_i \neq 0$ for some $i < n$. However, $\sum_{i<n} x_i x_i^* \neq 0$ whenever $x_i \neq 0$ for some $i < n$, where * is complex conjugation. Using this observation one easily obtains a sheaf theoretic version of the Gelfand-Naimark Theorem for C*-algebras $\mathcal{B} = \langle B, +, -, 0, \cdot, *, \bar{c}\rangle_{c \in C}$, where C is the set of complex numbers. The <u>Gelfand-Naimark Theorem</u> says that every commutative C*-algebra with identity is iseomorphic (that is isomorphic by a norm-preserving mapping) to the algebra C(I, C) of all continuous complex valued functions on a compact Hausdorff space I (See Simmons [34]). Globalizing we obtain:

<u>Theorem</u> 7.19 Every commutative C*-algebra with identity is isomorphic to a global subdirect product of local algebras over a compact Hausdorff space, where the quotient of each factor by its maximal ideal is isomorphic to the complex numbers. ∎

We doubt that anybody considers Theorem 7.19 an improvement over the Gelfand-Naimark Theorem. Of course in both Theorems 7.18 and 7.19 we started with a subdirect representation utilizing a natural topology on the factors, whereas the <u>discrete</u> sheaf construction induces the discrete topology on the stalks (Lemma 2.24). <u>Topological</u> sheaf constructions utilizing indiginous topologies on the stalks are not the topic of this study. The point we wish to make, however, is that even globalizing purely algebraic subdirect representations may yield results as peculiar as Theorem 7.18 and 7.19.

8. GLOBAL SUBDIRECT REPRESENTATION OF LATTICE ORDERED RINGS

The basic work on sheaf representation of lattice ordered rings is due to Keimel [24], and we shall frequently refer to the results of this paper by abbreviated quotation, such as "K3.3", which stands for Keimel [24, 3.3]. Since this paper is rather voluminous we have again selected a few central results to illustrate our methods. To establish the most general sheaf representation of lattice ordered rings we again only require a minute amount of ring theory. As to be expected, in general nothing tangible is known about the stalks of this representation. However, Keimel's main results are Boolean representations of several classes of function rings by totally ordered rings. Later Kennison [26] obtained the global representation of <u>all</u> function rings by totally ordered rings (Corollary 6.4). Thus, in retrospect, Keimel's results should be considered sharpened versions of Kennison's results for special classes of function rings, and we shall present them in this light.

Throughout this section we shall consider l-rings $\mathcal{B} = \langle B, +, -, 0, \cdot, \wedge, \vee \rangle$ as introduced in Section 6. l-rings are arithmetical and we immediately obtain a global subdirect representation from Theorem 4.42. This is essentially Wolf's [41] approach who claims to obtain Keimel's results from his version of Theorem 4.38. Since he does not elaborate to support his claim, we don't understand it. To the contrary, it appears to us that Keimel's approach is quite different and is actually more closely related to Hofmann's treatment of semi-prime rings with identity. To see what is involved we shall first review a few facts from Keimel [24]. So let $\mathcal{B}$ be an l-ring.

(1) (K1.4) $\langle B, \wedge, \vee \rangle$ is a distributive lattice.

(2) (K1.6) For $x \in B$ define

$$x^+ = x \vee 0; \quad x^- = (-x) \vee 0$$

Then $x = x^+ - x^-$.

Since an l-ring is congruence distributive, by Theorem 3.24 we may assume that $\mathcal{B} \subseteq \Pi\langle \mathfrak{A}_i \mid i \in I\rangle$ is subdirect, where $\{\theta_i^{\mathcal{B}} \mid i \in I\}$ is a set of irreducible congruences containing all completely meet irreducible congruences. Then $\mathcal{B}$ is fully expanded and, by Corollary 3.8, $\mathcal{B}$ induces a hull-kernel topology on I. For each congruence $\theta_J^{\mathcal{B}}$ define

$$M_J = \{x \in B \mid x\theta_J^{\mathcal{B}} 0\}$$

(3) The mapping $\theta_J^{\mathcal{L}} \to M_J$ is an isomorphism from the congruence lattice of $\mathcal{L}$ onto the lattice of convex ideals of $\mathcal{L}$.

Principal congruences on $\mathcal{L}$ correspond to principal convex ideals. $u \in B$ is called a formal unit if $\theta^{\mathcal{L}}(u, 0) = \mathbb{1}^{\mathcal{L}}$, that is u generates $\mathcal{L}$ as a convex ideal. For example, an identity is a formal unit. To give another example, if for each $x \in B$ there exists $n < \omega$ such that $nu > x$ then u is a formal unit.

The rudiments of patching come from

(4) (K1.11) Suppose for each $p < r$, M_p is a convex ideal of $\mathcal{L}$ and $b \in \sum_{p<r} M_p$, where $b \geq 0$. Then for each $p < r$ there exists $b_p \in M_p$ where $b_p \geq 0$ and $b = \bigvee_{p<r} b_p$.

Now we extract from Keimel's General Representation Theorem (K3.3) the basic algebraic fact that produces patching over the ring of hull-kernel closed sets.

Lemma 8.1 Let $\mathcal{L} \subseteq \Pi\langle \mathfrak{A}_i \mid i \in I\rangle$ be an l-ring, where $\{\theta_i^{\mathcal{L}} \mid i \in I\}$ is a set of irreducible congruences containing all completely meet irreducible congruences. Then $\mathcal{L}$ induces a hull-kernel topology on I and patches over the dual ring $\mathcal{R}$ of hull-kernel open sets.

Proof: Let $r < \omega$ and $F_p \in \mathcal{R}$, $x_p \in B$ for each $p < r$, where $I = \bigcup_{p<r} F_p$ and $F_p \cap F_q \subseteq E(x_p, x_q)$ whenever $p, q < r$. First notice that we may assume that $x_p \geq 0$ for each $p < r$. Indeed, if we can find $u, v \in B$ such that $F_p \subseteq E(u, x_p^+)$ and $F_p \subseteq E(v, x_p^-)$ for each $p < r$, then by (2), $F_p \subseteq E(u - v, x_p)$ for each $p < r$. By Lemma 3.29,

$$\mathbb{1}^{\mathcal{L}} = \bigvee_{p<r} \theta_{I-F_p}^{\mathcal{L}}$$

and by (3),

$$B = \sum_{p<r} M_{I-F_p}.$$

Let $b = \bigvee_{p<r} x_p$. By (4), there exist $b_p \in M_{I-F_p}$, for each $p < r$, such that $b_p \geq 0$ and $b = \bigvee_{p<r} b_p$. Now let $y = \bigvee_{p<r} x_p b_p$. We claim that $F_p \subseteq E(y, x_p)$ for each $p < r$. Let $i \in F_p$ and consider $q < r$. If $i \in F_q$ then $x_q(i) = x_p(i)$ by hypothesis. Otherwise $i \in I - F_q$ and $b_q(i) = 0$ since $b_q \in M_{I-F_q}$. In either case

$$x_q(i) \wedge b_q(i) = x_p(i) \wedge b_q(i)$$

so that by (1),

$$y(i) = x_p(i) \wedge \bigvee_{p<r} b_q(i) = x_p(i) \wedge b(i) = x_p(i). \blacksquare$$

Notice that Lemma 8.1 is a perfect analogue of Lemma 7.2. Next we obtain an analogue of Theorem 7.4.

<u>Theorem</u> 8.2 Let $\mathcal{B} \subseteq \Pi\langle \mathfrak{A}_i \mid i \in I\rangle$ be an l-ring, where $\{\theta_i^{\mathcal{B}} \mid i \in I\}$ is the set of all irreducible (completely meet irreducible) congruences on $\mathcal{B}$. Then $\mathcal{B}$ is isomorphic to a global subdirect product

$$\mathcal{B} \cong g(\mathcal{B}) \subseteq \Pi\langle \mathcal{B}/\mathcal{U} \mid \mathcal{U} \in J\rangle,$$

where J is the set of ultrafilters on the ring of hull-kernel closed subsets of I.

<u>Proof</u>: Use Theorem 4.13 and Lemma 8.1. $\blacksquare$

As usual, in general we don't know anything more about the factors $\mathcal{B}/\mathcal{U}$ than we know about $\mathcal{B}$. To obtain more specific information about the factors Keimel restricts his attention to function rings and we follow suit. (See Section 6). Keimel calls an f-ring <u>projectable</u> if the pseudo complement of every principal congruence is a factor congruence (Notice that the definition of this notion requires a distributive congruence lattice!) To investigate this notion we represent an f-ring $\mathcal{B}$ as a subdirect product $\mathcal{B} \subseteq \Pi\langle \mathfrak{A}_i \mid i \in I\rangle$, where $\{\theta_i^{\mathcal{B}} \mid i \in I\}$ is the set of <u>all</u> irreducible congruences on $\mathcal{B}$.

<u>Lemma</u> 8.3 $\mathcal{B}$ is projectable if and only if $\overline{D(x, y)}$ is clopen for every $x, y \in B$.

<u>Proof</u>: By Lemma 5.20, for every $x, y \in B$

$$\theta^{\mathcal{B}}(x, y)^{\sim} = \theta^{\mathcal{B}}_{D(x, y)} = \theta^{\mathcal{B}}_{\overline{D(x, y)}} .$$

By Lemma 4.43, the mapping $F \to \theta_F^{\mathcal{B}}$ is an anti isomorphism from the field of clopen subsets of I onto the Boolean algebra of factor congruences of $\mathcal{B}$. $\blacksquare$

We shall now pass to the space of <u>minimal irreducible</u> congruences. In any algebra, by Zorn's Lemma, every irreducible congruence contains a minimal irreducible congruence. Thus, if $\{\theta_i^{\mathcal{B}} \mid i \in K\}$ is the set of minimal irreducible congruences on $\mathcal{B}$ then

$$\mathcal{B} \cong \pi_K(\mathcal{B}) \subseteq \Pi\langle \mathfrak{A}_i \mid i \in K\rangle$$

is still a subdirect representation of $\mathcal{B}$. If $i \in I$ and $k \in K$, where $\theta_k^{\mathcal{B}} \subseteq \theta_i^{\mathcal{B}}$, then $i \in \overline{\{k\}}$ so that

$$I = \bigcup\{\overline{\{k\}} \mid k \in K\}$$

and K is dense in I. We need one more algebraic fact from Keimel [24].

<u>Lemma</u> 8.4 If $x, y \in B$ then

$$K \cap E(x, y) \cap \overline{D(x, y)} = \emptyset.$$

<u>Proof</u>: Suppose $i \in E(x, y) \cap \overline{D(x, y)}$. Then

$$\theta^{\mathcal{B}}_{E(x, y)} = \theta^{\mathcal{B}}(x, y) = \theta^{\mathcal{B}}(x - y, 0) \subseteq \theta_i^{\mathcal{B}}$$

and

$$\theta^{\mathcal{B}}_{\overline{D(x, y)}} = \theta^{\mathcal{B}}(x, y)^{\sim} = \theta^{\mathcal{B}}(x - y, 0)^{\sim} \subseteq \theta_i^{\mathcal{B}}.$$

In terms of convex ideals this means that $\langle x - y\rangle \subseteq M_i$ and $\langle x - y\rangle^{\sim} \subseteq M_i$. By K6.5a, M_i is not a minimal irreducible ideal and therefore $i \notin K$. ∎

<u>Theorem</u> 8.5 Let $\mathcal{B}$ be an f-ring. If $\mathcal{B} \subseteq \Pi\langle \mathfrak{A}_i \mid i \in I\rangle$ is subdirect, where $\{\theta_i^{\mathcal{B}} \mid i \in I\}$ is the set of all irreducible congruences on $\mathcal{B}$, and $K \subseteq I$, where $\{\theta_i^{\mathcal{B}} \mid i \in K\}$ is the set of minimal irreducible congruences on $\mathcal{B}$ then

(i) $\mathcal{B} \cong \pi_K(\mathcal{B}) \subseteq \Pi\langle \mathfrak{A}_i \mid i \in K\rangle$;

(ii) $\mathcal{B}$ is projectable if and only if $\pi_K(\mathcal{B})$ is closed under $\bar{n}$.

<u>Proof</u>: (ii) Suppose $\mathcal{B}$ is projectable and let $x, y, u, v \in B$. By Lemma 8.3, $\overline{D(x, y)}$ is clopen and by Lemma 8.1, there exists $z \in B$ with $\overline{D(x, y)} \subseteq E(z, v)$ and $I - \overline{D(x, y)} \subseteq E(z, u)$. Now

$$D(\pi_K(x), \pi_K(y)) = D(x, y) \cap K \subseteq \overline{D(x, y)}$$

and by Lemma 8.4,

$$E(\pi_K(x), \pi_K(y)) = E(x, y) \cap K \subseteq I - \overline{D(x, y)}.$$

Thus

$$\pi_K(z) = \bar{n}(\pi_K(x), \pi_K(y), \pi_K(u), \pi_K(v)) \in \pi_K(B).$$

Conversely, suppose $\pi_K(\mathcal{B})$ is closed under $\bar{n}$. We shall show that $\overline{D(x, y)}$ is open. Let $i \in \overline{D(x, y)}$. Choose $v \in B$ with $i \in D(v, 0)$, and let $z \in B$ where

$$\pi_K(z) = \bar{n}(\pi_K(x), \pi_K(y), \pi_K(0), \pi_K(v)).$$

Then

$$E(x, y) \cap K \subseteq E(0, z)$$
$$D(x, y) \cap K \subseteq E(z, v).$$

We claim that $i \in D(z, 0) \subseteq \overline{D(x, y)}$. Indeed, since each point of $D(x, y)$ must be in the closure of a point in $D(x, y) \cap K$, we have

$$i \in \overline{D(x, y)} = \overline{D(x, y) \cap K}.$$

Thus $0 \neq v(i) = z(i)$, so that $i \in D(z, 0)$. Now suppose $j \notin \overline{D(x, y)}$. There exists $k \in K$ with $j \in \overline{\{k\}}$ and therefore $k \notin \overline{D(x, y)}$. By Lemma 8.4, $k \in K \cap E(x, y)$ and therefore $z(k) = 0$. It follows that $z(j) = 0$, that is $j \in E(z, 0)$. This establishes our claim and $\overline{D(x, y)}$ is open. ∎

Theorem 8.6 (Keimel) An f-ring is projectable if and only if it is Boolean representable by totally ordered rings.

Proof: If $\mathcal{B}$ is a projectable f-ring, use Theorem 8.5 and Corollary 5.11. Since the totally ordered rings are exactly the subdirectly indecomposable f-rings, all "old" factors are totally ordered. Since the class of totally ordered rings is universal, all "new" factors are totally ordered. Finally, use Theorem 5.31 to get rid of repeated trivial factors. This shows that $\mathcal{B}$ is Boolean representable by totally ordered rings.

For completeness we give Keimel's proof of the converse. Suppose $\mathcal{B} \subseteq \Pi\langle \mathfrak{A}_i \mid i \in I\rangle$ is Boolean subdirect, where each $\mathfrak{A}_i$ is totally ordered. Consider $x \in B$. Since $\mathcal{B}$ is closed under $\bar{n}$, $\theta^{\mathcal{B}}_{E(x, 0)}$ and $\theta^{\mathcal{B}}_{D(x, 0)}$ are complementary. We claim that $\theta^{\mathcal{B}}(x, 0)^{\sim} = \theta^{\mathcal{B}}_{D(x, 0)}$. In terms of l-ideals we must show that

$$\langle x\rangle^{\sim} = M,$$

where $\langle x\rangle$ is the principal l-ideal generated by x and

$$M = \{y \in B \mid D(x, 0) \cap D(y, 0) = \emptyset\}.$$

Now clearly $\langle x\rangle \cap M = \{0\}$, and M is an l-ideal. If $y \in \langle x\rangle^{\sim}$ then

$$|x| \wedge |y| = (xv - x) \wedge (yv - y) \in \langle x\rangle \cap \langle x\rangle^{\sim} = \{0\}.$$

For each $i \in I$, $|x(i)| \wedge |y(i)| = 0(i)$ and, since $\mathfrak{A}_i$ is totally ordered, $x(i) = 0(i)$ or $y(i) = 0(i)$. Thus $y \in M$. ∎

As we mentioned in Section 6, the totally ordered rings are exactly the subdirectly indecomposable f-rings. Apparently no intrinsic ring theoretic characterization of the subdirectly irreducible f-rings is known. Nevertheless, in view of Theorem 8.6 we may ask (compare Problem 6.19):

Problem 8.7 Which f-rings are Boolean representable?

A still more special question was answered by Keimel [24]:

Theorem 8.8 An f-ring is quasiregular if and only if it is Boolean representable by simple (totally ordered) f-rings.

In fact, Theorem 8.8 follows from a more general result:

Theorem 8.9 An l-ring is quasiregular if and only if it is Boolean representable by simple l-rings.

Proof: Use Theorem 5.24. ∎

Our last application to f-rings corresponds to K4.9, where Keimel claims that every l-ring is isomorphic to a global subdirect product of directly indecomposable l-rings over a Boolean space. Now in general this claim is not correct, however Keimel has pointed out to us that his result does hold for f-rings with a formal unit. Actually, Keimel's representation is the same as our Theorem 4.45, and by Lemma 4.50 the factors of this representation are directly indecomposable if and only if the index space has the component separation property. The crucial issue therefore is the component separation property. We first need an auxiliary fact.

Lemma 8.10 Let $\mathfrak{B} \subseteq \Pi\langle \mathfrak{A}_i \mid i \in I\rangle$ be an f-ring, where $\{\theta_i^{\mathfrak{B}} \mid i \in I\}$ is the set of completely meet irreducible congruences on $\mathfrak{B}$. Then the following are equivalent:

(i) $\mathfrak{B}$ has a formal unit.

(ii) $I = D(x, y)$ for some $x, y \in B$.

(iii) B is a finitely generated convex ideal.

(iv) $\mathbb{1}^{\mathfrak{B}}$ is a compact congruence.

Proof: u is a formal unit if and only if it is not contained in any proper convex ideal, if and only if $D(u, 0) = I$. This proves the equivalence of (i) and (ii), and the equivalence of (i) and (iii) is K1.20. Finally, (iv) holds if and only if $\mathbb{1}^{\mathfrak{B}}$ is a finite join of principal congruences, if and only if (iii) holds. ∎

Now let $\mathcal{L}$ be an f-ring and let $\operatorname{Irrd}\Theta(\mathcal{L})$ and $\operatorname{Max}\Theta(\mathcal{L})$ be the sets of irreducible and maximal congruences on $\mathcal{L}$ respectively. We quote the pivotal algebraic facts from Keimel:

<u>Lemma</u> 8.11 Let $\mathcal{L}$ be an f-ring

(i) (K5.13) Each irreducible congruence ψ is contained in a unique maximal congruence $\alpha(\psi)$.

(ii) (K8.11) If $\mathcal{L}$ has a formal unit then $\operatorname{Max}\Theta(\mathcal{L}) \subseteq \operatorname{Irrd}\Theta(\mathcal{L})$ and $\alpha:\operatorname{Irrd}\Theta(\mathcal{L}) \to \operatorname{Max}\Theta(\mathcal{L})$ is a continuous retraction.

(iii) (K5.17) $\operatorname{Max}\Theta(\mathcal{L})$ is a Hausdorff space. ∎

<u>Lemma</u> 8.12 If $\mathcal{L}$ is an f-ring with a formal unit then $\operatorname{Max}\Theta(\mathcal{L})$ is compact.

<u>Proof</u>: We use Lemma 3.28. Let $\mathcal{L} \subseteq \Pi\langle \mathfrak{A}_i \mid i \in I\rangle$, where $\{\theta_i^{\mathcal{L}} \mid i \in I\} = \operatorname{Irrd}\Theta(\mathcal{L})$, and let $M \subseteq I$, where $\{\theta_i^{\mathcal{L}} \mid i \in M\} = \operatorname{Max}\Theta(\mathcal{L})$. Then $\pi_M(\mathcal{L})$ induces a hull-kernel topology on M (which is the relative topology). Since $\mathcal{L}$ has a formal unit, so does $\pi_M(\mathcal{L})$ and, by Lemma 8.10, $\mathbb{1}^{\pi_M(\mathcal{L})}$ is compact. Finally, $\{\theta_i^{\pi_M(\mathcal{L})} \mid i \in M\}$ is the set of maximal congruences on $\pi_M(\mathcal{L})$ so that Lemma 3.28 applies. ∎

<u>Theorem</u> 8.13 If $\mathcal{L}$ is an f-ring with a formal unit then $\mathcal{L}$ is isomorphic to a global subdirect product of directly indecomposable f-rings over a Boolean space.

<u>Proof</u>: To apply Theorem 4.51 we have to verify that $\operatorname{Irrd}\Theta(\mathcal{L})$ has the component separation property. By Lemmas 8.11 and 8.12, $\operatorname{Max}\Theta(\mathcal{L})$ is a compact Hausdorff space. It is well-known that a compact Hausdorff space has the component separation property (see e.g., Hocking and Young [8]), and the desired conclusion follows from Lemma 8.11 (ii). ∎

In view of Kennison's result (Corollary 6.4) Theorem 8.13 is of rather limited interest, because we do not obtain a Boolean subdirect representation (equalizers are not necessarily clopen) and the Boolean space is obtained at the expense of a further proliferation of factors. For example, if I is any connected space then the ring $C(I, R)$ of all continuous real valued functions on I is a directly indecomposable f-ring with formal unit. Such a ring is clearly not totally ordered. In our view this once again demonstrates the need for a relentless inquiry into the nature of the factors in sheaf representation.

REFERENCES

1. Arens, R. F. and Kaplansky, I., Topological representation of algebras, Trans. Amer. Math. Soc. 63(1968), 457-481.

2. Balbes, R., and Dwinger, P., Distributive Lattices, University of Missouri Press, Columbia, Missouri (1974).

3. Birkhoff, G., and Pierce, R. S., Lattice-ordered rings, Anais Acad. Brasil. Ci. 28 (1956), 41-69.

4. Bullman-Fleming, S., and Werner, H., Equational compactness of quasi primal varieties (Preprint, 1975).

5. Burris, S., and Werner, H., Sheaf constructions and their elementary properties, Trans. Amer. Math. Soc. (to appear).

6. Burris, S., Sub-Boolean power representations in quasi-primal generated varieties, Preprint.

7. Clifford, A. H., Partially ordered Abelian groups, Annals Math. 41(1940), 465-473.

8. Comer, S., Representations of algebras by sections over Boolean spaces, Pacific J. of Math. 38(1971), 29-38.

9. Comer, S. D., Restricted direct products and sectional representations, Math. Nachrichten 64(1974), 333-344.

10. Comer, S. D., Elementary properties of structures of sections, Boletin de la Sociedad Matematica Mexicana 19(1974), 78-85.

11. Dauns, J., and Hofmann, K. H., The representation of biregular rings by sheaves, Math. Z. 91(1966), 103-123.

12. Dauns, J., and Hofmann, K. H., Representation of rings by sections, Memoirs Amer. Math. Soc. 83(1968).

13. Davey, B. A., Sheaf spaces and sheaves of universal algebras, Math. Z. 134(1973), 275-290.

14. Fuchs, L., Partially Ordered Algebraic Systems, Pergamon Press (1963)

15. Fried, E., and Pixley, A. F., The dual discriminator function in universal algebra, Preprint.

16. Gould, M. I., and Gratzer, G., Boolean extensions and normal subdirect powers of finite universal algebras, Math. Z. 99(1967), 16-25.

17. Gratzer, G., Universal Algebra, D. Van Nostrand (Toronto,1968).

18. Hashimoto, J., Direct, subdirect decompositions and congruence relations, Osaka J. Math. 9(1957), 87-112.

19. Hocking, J. G., and Young G. S., Topology, Addison-Wesley (Reading, Mass., 1961).

20. Hofmann, K. H., Representation of algebras by continuous sections, Bulletin Amer. Math. Soc. 78(1972), 291-373.

21. Jacobson, N., A topology for the set of primitive ideals in an arbitrary ring, Proc. Nat. Acad. Sci. 31(1945), 333-338.

22. Jacobson, N., Structure of Rings, Amer. Math. Soc. Colloq. Publications, Vol. XXXVII (Providence, 1968).

23. Jonsson, B., Algebras whose congruence lattices are distributive, Math. Scand. 21(1967), 110-121.

24. Keimel, K., The representation of lattice ordered groups and rings by sections in sheaves, Lecture Notes in Math. 248. Springer Verlag (Berlin and New York, 1971).

25. Keimel, K., and Werner, H., Stone duality for varieties generated by quasi primal algebras, Memoirs Amer. Math. Soc. 148(1974), 59-85.

26. Kennison, J.F., Integral domain type representations in sheaves and other topoi, Math. Z. 151(1976), 35-56.

27. Krauss, P., On quasi primal algebras, Math. Z. 134(1973), 85-89.

28. Lambek, J., Lectures on Rings and Modules, Blaisdell Publ. Co. (London, 1966).

29. Ledbetter, C. S., A criterion for the representability of algebras by sections in sheaves, Notices Amer. Math. Soc. 24(1977), A 463.

30. Macintyre, A., Model-completeness for sheaves of structures, Fund. Math. 81(1973), 73-89.

31. Pierce, R. S., Modules over commutative regular rings, Memoirs Amer. Math. Soc. 70(1967).

32. Pierce, R. S., Introduction to the Theory of Abstract Algebra, Holt, Reinhart and Winston (New York, 1968).

33. Pixley, A. F., Completeness in arithmetical algebras, Alg. Univ. 2(1972), 172-196.

34. Simmons, G.F., Topology and Modern Analysis, McGraw-Hill (New York, 1963).

35. Stone, M. H., A general theory of spectra I and II, Proc. Nat. Acad. Sci. USA 26(1940), 280-283, and 27(1941), 83-87.

36. Wallman, H., Lattices and topological spaces, Annals of Math. 39(1938), 112-126.

37. Weispfenning, V., Model-completeness and elimination of quantifiers for subdirect products of structures, J. of Algebra 36(1975), 252-277.

38. Weispfenning, V., Lattice products - a model theoretic approach to sheaves, Preprint.

39. Werner, H., Algebraic representation and model theoretic properties of algebras with the ternary discriminator (Preprint, 1975).

40. Werner, H., A generalization of Comer's sheaf representation theorem, Preprint.

41. Wolf, A., Sheaf representation of arithmetical algebras, Memoirs Amer. Math. Soc. 148(1974), 87-93.

GhK, OE 3
3500 Kassel
Heinrich-Plett-Str. 40
Fed. Rep. of Germany

SUNY, New Paltz
New Paltz, N.Y. 12561
U S A